Wireless METHODS

Wireless METHΘDS

Wireless Network Dimensioning and Strategy Development through Computational Techniques

Derek McAvoy

First paperback edition March 2021

Edited by Peter Wolchak
Book design by Kathleen Gray

ISBN 978-1-7774641-0-3 (paperback)

Contents

1 Introduction

Most mathematical books on wireless communications focus on the physical layer. They explain the probabilistic and statistical techniques that are used to model the physical RF channel which is known to distort the signal, and then develop optimal modulation and coding schemes to mitigate the impacts of distortion and noise. Mobile network operators (MNOs) rely on device and network equipment vendors to supply smartphones and base station radio equipment built to industry standards designed to address the challenges of the RF physical communication layer. But once commercial equipment becomes available, the problem shifts from the physical layer to the development of optimal network-deployment strategies. This higher-layer analysis is the responsibility of the MNO and the focus of this book.

The primary objective of an MNO is to build a wireless network that can support the expected amount of traffic while satisfying a number of Quality of Experience, coverage, economic and regulatory goals. Although equipment vendors will provide some assistance, it is the responsibility of the MNO to develop a method for network dimensioning and wireless strategy development. At a minimum, all methods should incorporate data collected from the wireless network. Called Operational Measurements (OMs), these provide insight into various key performance indicators (KPIs) such as the amount of traffic, level of utilization, and perceived user throughput for each base-station radio.

This book will outline several approaches to dimensioning wireless networks. The first is a non-rigorous approach based on examining Operational Measurements (OMs) from base-station radios and employing some simple heuristics. The method is easy to explain, and someone taught it would have great difficulty finding reasons not to use it. One of its weaknesses is the inability to estimate the improvement in capacity provided by new wireless features. This estimate must be provided by either measurements from field trials or from the results of Monte Carlo simulations.

Is Monte Carlo simulation suitable for network planning or wireless strategy development? Most Monte Carlo simulators used in the industry require days if not weeks of computing time to simulate a single scenario of a network consisting of only several dozen cells. Most wireless networks consist of several thousand cells with a wide range of configurations that change with time. Monte Carlo simulation is suitable to test the functionality of new protocols and wireless techniques but, for network planning and developing wireless strategies, it is much too cumbersome to provide the necessary answers in the required time frame.

Ideally, an analytical approach would be ideal; that is an equation or a deterministic algorithm using a finite number of steps to arrive at the answer. My search for an analytical answer was impeded when I encountered calculations I believed to be intractable using a deterministic approach. Fortunately, those calculations succumbed to a stochastic approach, the same one used in Monte Carlo simulations. The techniques that are the foundation of Monte Carlo simulations—which have long been used in the development of many physical-layer protocols—are now being exploited for higher-layer network analysis.

This led, after decades of research, to another method that uses stochastics to expedite the development of wireless strategies and is the main focus of this book.

If you are employed in the wireless industry, you may encounter any of the following statements:

- The sole purpose of acquiring spectrum is to increase network capacity.
- Adding cell sites always increases network capacity.
- More complicated wireless technology features are always more valuable than simple ones.
- Wireless network planning is very time consuming and difficult.
- The lower the path-loss exponent, the better the network performance.
- Network capacity is greatly influenced by the design of the scheduler used by the base-station radio.
- Wireless Priority Service is a very expensive feature to implement for the mobile network operator (MNO).
- Low-frequency spectrum bands, due to their lower RF path loss, are always more valuable than high-frequency spectrum bands.
- Every macro cell sector has traffic hotspots, so it makes sense to find those hotspots and place a small cell to offload traffic.
- A network using MU-MIMO (multi-user multiple-input and multiple-output)—which is an advanced antenna system capable of serving multiple subscribers simultaneously—will always provide higher network capacity than a network using simple beamforming antennas which only can transmit data to one subscriber at a time.
- It is always the case that when the radios in a wireless network transmit more often, the amount of wireless traffic will be greater.

The validity of the above statements, considered by many as self-evident, can be tested using the material in this book. For those who do not wish to read the entire book to find out which ones are truly valid, you can skip to the concluding chapter.

MNOs are particularly interested in a quantitative answer to the following questions:

- How much does it cost to transport a gigabyte of data?
  - What factors influence the cost?
- How much money should be spent to acquire a spectrum licence?
  - What impact does the frequency band have on spectrum value?
  - What other factors affect the value of the spectrum licence?
- What are the capacity gains associated with a new feature?
- How fast are data speeds (average and lowest 5th percentile) during the busiest hour of the day?
  - How do those speeds change when more traffic is loaded onto the network?
- What is the best method to implement a new technology?
- Can the wireless network support the data speeds and capacity requirements of a new service?

The method I proposed can easily answer those questions. Once you understand the techniques in this book, there is no longer any need to rely on the words of so-called experts. No advanced degree in mathematics or detailed knowledge of machine-learning techniques are required to understand the algorithms developed in this book. All you need is high-school-level math and some basic MATLAB programming skills. Some rudimentary statistics are used to validate different methods of dimensioning wireless networks, but none are required to explain how these methods work.

A quick note about MATLAB, for those unfamiliar with the software: MATLAB is desktop software designed for iterative analysis and design processes; it also offers a programming language that directly expresses matrix and array mathematics. It's an environment that works for me, but there are many other programming languages that can be used, such as R and Python.

In the first chapter, I will describe how an MNO may perform network dimensioning and strategic planning. I compare that method with an alternative that I will more fully develop in subsequent chapters.

The second chapter develops the necessary underlying theory used in the development of the tool. My explanation of the theorems might be considered cursory, but I find too often that simple concepts become obscured for the sake of comprehensiveness.

The third chapter is the meat of the book, in which I develop a stochastic algorithm to perform calculations that are otherwise intractable. Although I used some advanced mathematical concepts in developing the model, the reader does not need to know any of them. Increasingly, calculating the solution of a set of equations are performed by computer algorithms. To explain how the model works, MATLAB code snippets are used to analyze the performance of a simple toy (experimental) network. Despite the simplicity of the toy network, the tool was able to generate insights that I did not intuit.

In the fourth chapter, I use the stochastic approach developed in the previous chapter to analyze two LTE-A features: Carrier Aggregation and Coordinated Multipoint. An attempt is made to find the value of both Carrier Aggregation and Coordinated Multipoint, using the insights garnered in the previous chapter. Investigating the value of Carrier Aggregation also gives you a better understanding of the value of spectrum.

The fifth chapter extends the tool to analyze larger networks with multi-sectored cell sites. The main goal here is to determine the value of different spectrum bands in various morphologies (i.e. rural, suburban, urban, dense urban). This chapter also reveals the trade-offs between coverage and capacity.

The sixth chapter examines small-cell deployment issues. The stochastic approach explains why certain small-cell design practice guidelines are far from optimal and suggests solutions to increase their effectiveness by an order of magnitude.

The seventh chapter analyzes an early implementation of a promising technology called MU-MIMO, comparing its performance with current network designs and to a related but simpler technology called adaptive beamforming.

You may wonder what makes this book different from the multitude covering general wireless communication theory or books examining features of various standards-based air-interface technologies such as LTE and HSPA. Notably, those books are written by researchers working in academia or employed by wireless network equipment vendors. In contrast, my background includes working for two major mobile network operators in various wireless technology research, development and strategy departments over a period spanning three decades. If the contents of the book are supposedly based on scientific analysis, why is the background of the author relevant? There is a famous quote by John Tukey, an American mathematician and statistician who coined the term "bit" for the binary digit and was the first person to use the term "software" in print:

> *Far better an approximate answer to the right question, which is often vague, than an exact answer to the wrong question, which can always be made precise.*

Most papers that evaluate the performance of LTE and HSPA focus on maximizing the amount of network data traffic by operating every base-station radio at 100% utilization. However, MNOs never operate the wireless network in this state due to A) the severe impact it has on customer experience, and B) traffic not being uniformly distributed as assumed by most papers on LTE and HSPA capacity. It is well known that actual wireless networks rarely exceed 30% utilization. Indifferent to Tukey's admonition, after spending a considerable amount of effort to precisely calculate the spectral efficiency of LTE under universal full-buffer congestion, researchers doubled down on using this improbable network condition in search of novel schedulers to increase LTE capacity which are only effective when data buffers are full. Few operators are searching for techniques to improve network performance in a scenario that most would deem as unrealistic.

This book will not go into the derivation of numerous formulas in excruciating detail or describe the intricacies of assorted features of current air-interface standards. There are many textbooks that adequately cover that material. The results in this book could be criticized because I ignored various aspects of the wireless protocol in order to reduce the time to perform analysis and sources of errors. Faulting the excessive use of abstraction is not without merit, but the benefits far outweigh the disadvantages. Referring to Tukey's quote, the goal is not to provide answers with unparalleled levels of accuracy, but a method to help guide MNOs in developing wireless strategies. I will leave it to the reader to decide whether my design choices are reasonable.

Although this book is written for those who work for network operators, it will also benefit those employed by equipment vendors or in academia. I often wonder if equipment vendors search for novel solutions and then hunt for data to verify efficacy. The expectation is that any design enhancement will provide some performance improvement, often unspecified. I have nothing against trial and error, however, without a computer algorithm to evaluate the efficacy of a new technique you are essentially working in a fog. Often network features do not provide the touted benefits while others can be made much more effective if preconceived notions of how to deploy them are ignored. At what point do vendors optimize their solution beyond the point of diminishing returns? I'm advocating that there is as much benefit in researching better methods in evaluating wireless technologies as devising novel techniques which appear efficacious but are impossible to analyze without the assistance of a computer algorithm.

I also understand that not everyone will want to write MATLAB programs, dimension wireless networks or develop wireless strategies. I've written this book to give some insight into the nature of wireless strategy development from the perspective of an MNO. The material will give the reader a high level of understanding of some of the major scientific and mathematical discoveries that influenced the development of wide-area wireless networks. Perhaps with this understanding, the reader will embark on discovering other methods of planning wireless networks.

1 | Network Dimensioning

Dimensioning a wireless network involves determining the number of cell sites required to meet a Quality of Experience target, given the level of traffic during the busiest period of the day and the bandwidth of the available spectrum. This exercise is necessary to understand the cost of providing wireless service. There are various approaches that can be used to dimension a network, ranging from simple heuristics to full-blown Monte Carlo simulations.

I will assume that the reader is familiar with the basic concepts of cellular networking. A base station consists of one or more sectors, each capable of supporting a certain amount of data traffic, typically measured in Mbps. All vendors of mobile wireless networking equipment provide Operational Measurements (OMs) which can be configured to, among other things, determine the amount of traffic each radio in the network carries for each hour in the day. The hour of greatest interest is the busiest hour of the day, i.e. the hour where the network carries the most traffic. An engineer in the Operations and Engineering Planning department may download busy-hour traffic statistics for two particular clusters of sectors; an example is shown in the table below. All analyses in this chapter assume that traffic statistics are associated with the downlink (the link where data is transmitted from the base-station radio to the mobile receiver) since it is typically much busier than the uplink.

Table 1: *Downlink busy-hour sector traffic for a rural and urban cluster.*

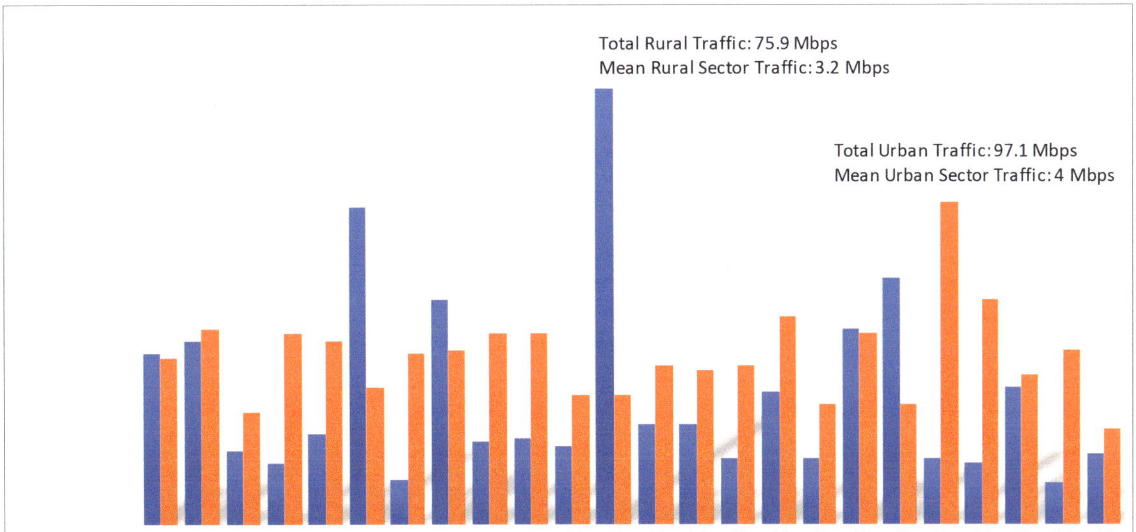

Sector No.	1	2	3	4	5	6	7	8	9	10	11	12	13	14	15	16	17	18	19	20	21	22	23	24
Rural (Mbps)	4.1	4.4	1.8	1.5	2.2	7.6	1.1	5.4	2.0	2.1	1.9	10.9	2.4	2.4	1.6	3.2	1.6	4.7	5.9	1.6	1.5	3.3	1.0	1.7
Urban (Mbps)	4.0	4.7	2.7	4.6	4.4	3.3	4.1	4.2	4.6	4.6	3.1	3.1	3.8	3.7	3.8	5.0	2.9	4.6	2.9	7.7	5.4	3.6	4.2	2.3

Suppose another operational measurement statistic indicates that sector 12 in the rural cluster and sector 20 in the urban cluster are at 100% resource utilization and unable to carry more traffic. If those sectors are fully utilized and data traffic is forecasted to grow, the network planner in charge of dimensioning may advise that a new cell site be deployed to offload traffic from those congested sectors.

Assume that the wireless technology is 10 MHz of LTE, and the base-station radio and mobile receiver support 2×2 MIMO. There are several points worth noting:

1. Industry papers indicate that 10 MHz of LTE should be able to support 15 Mbps of traffic on average.[1] That capacity figure is just an estimate obtained from network simulations, and there are no guarantees that LTE radios deployed in an actual network will meet those figures. Since network configurations, RF path-loss assumptions and radio parameters used in simulations will differ from those found in actual networks, it is more likely that the engineer responsible for network planning will trust the capacity figures collected from OMs. In this case, he or she may presume that a rural sector should be able to support 10.9 Mbps and an urban sector can only support 7.7 Mbps.
2. The average traffic for sectors in the rural cluster is lower than the average traffic in the urban cluster. This is because there are many coverage cell sites in a rural morphology that do not carry much traffic, and very few busy macro sectors covering, for example, a small town. The traffic levels of urban cell sites are more uniform as new cell sites are constantly added to offload busy sectors. (See Figure 1.)

Figure 1: *Chart of sector traffic for a rural and urban cluster.*

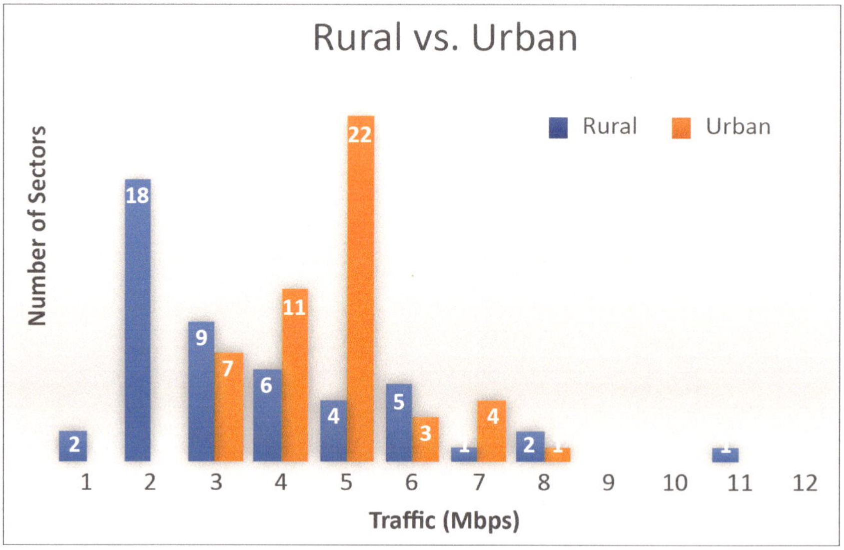

1 From "HSPA to LTE-Advanced: 3GPP Broadband Evolution to IMTAdvanced (4G)", Peter Rysavy, *Rysavy Research*, September 2009.

3. Traffic was quoted in Mbps. A figure of merit called "spectral efficiency" is the traffic at full capacity divided by the spectrum bandwidth for a given link. For example, the spectral efficiency of the busiest sector of the rural cluster is 10.9 Mbps / 10 MHz = 1.09 bps/Hz. For the urban morphology, the spectral efficiency is only 0.77 bps/Hz. Of greater interest is how many GB the network can support in a given month, since that will determine how many subscribers can be supported. I will show later that to convert spectral efficiency in bps/Hz to GB/Month/MHz, one needs to multiply by 100. That 1.09 bps/Hz is equivalent to 109 GB/Month/MHz.
4. It is curious that in the above example, the rural sector at full capacity (rural sector 12) carries more traffic than the urban sector at full capacity (urban sector 20). Intuition would suggest it is more likely that a mobile subscriber in a rural area would be farther from a cell site, receive a weaker signal and experience relatively lower data speeds than a mobile subscriber in an urban area. It is counterintuitive that rural cell sites are able to carry more traffic than can urban ones, but this situation often happens in actual networks. We will revisit this mystery throughout the book.

One reason for analyzing sector traffic is to determine which sectors require traffic offloading when network traffic increases. Typically, the marketing department will inform network planners of the forecasted growth for a number of years—as many as 10 years. Assume that marketing predicts that traffic will grow 40% per year. The table below shows the traffic for the rural cluster after one year.

Table 2: *Busy-hour sector traffic for rural cluster.*

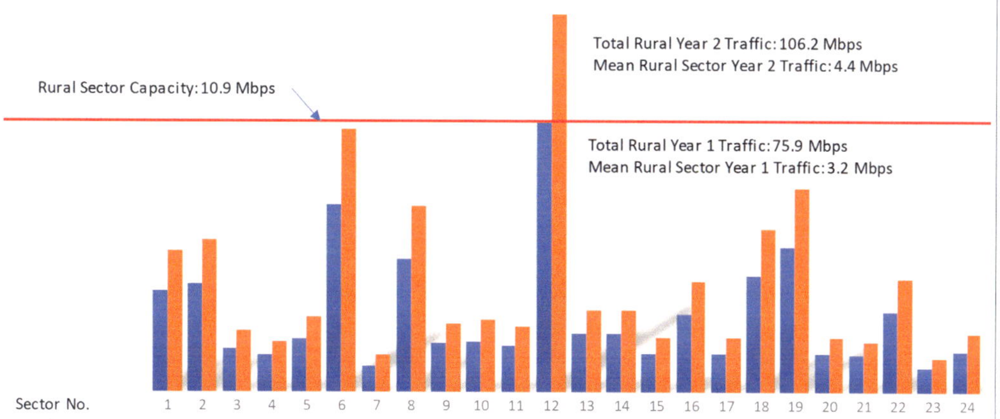

Sector No.	1	2	3	4	5	6	7	8	9	10	11	12	13	14	15	16	17	18	19	20	21	22	23	24
Rural Year 1 (Mbps)	4.1	4.4	1.8	1.5	2.2	7.6	1.1	5.4	2.0	2.1	1.9	10.9	2.4	2.4	1.6	3.2	1.6	4.7	5.9	1.6	1.5	3.3	1.0	1.7
Rural Year 2 (Mbps)	5.7	6.2	2.5	2.1	3.1	10.6	1.5	7.6	2.8	2.9	2.7	15.3	3.4	3.4	2.2	4.5	2.2	6.6	8.3	2.2	2.1	4.6	1.4	2.4

Note that in year 2, the traffic in sector 12 has exceeded the assumed capacity of a rural site of 10.9 Mbps. This means that a new cell site is required to offload traffic from sector 12. When adding a cell site, the planner has to estimate how much traffic to offload from neighboring sectors. Since this exercise must be done for thousands of cell sites for a period extending sometimes up to 10 years, shortcuts are often taken. For example, one may assume that the new cell site only offloads traffic from the congested sector. Since most cell sites have three sectors that carry different amounts of traffic, the planner must arbitrarily determine how much traffic each sector of the new cell site will transmit. In this example, the network planner assumes that the new cell site will offload 50% of the traffic from the congested sector. The amount of traffic for each sector of the new cell site will be randomly assigned, but the sum of the traffic will be equal to the amount of traffic offloaded from the congested sector.

Table 3:
Busy-hour sector traffic for rural cluster with added three-sectored cell site.

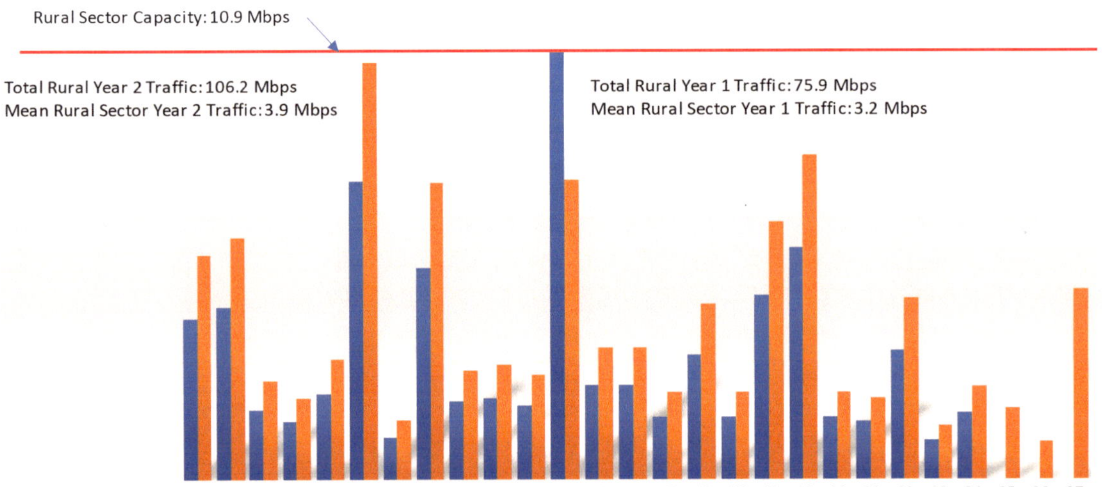

Sector No.	1	2	3	4	5	6	7	8	9	10	11	12	13	14	15	16	17	18	19	20	21	22	23	24	25	26	27
Rural Year 1 (Mbps)	4.1	4.4	1.8	1.5	2.2	7.6	1.1	5.4	2.0	2.1	1.9	10.9	2.4	2.4	1.6	3.2	1.6	4.7	5.9	1.6	1.5	3.3	1.0	1.7			
Rural Year 2 (Mbps)	5.7	6.2	2.5	2.1	3.1	10.6	1.5	7.6	2.8	2.9	2.7	7.6	3.4	3.4	2.2	4.5	2.2	6.6	8.3	2.2	2.1	4.6	1.4	2.4	1.8	1.0	4.8

Note that for year 2, the traffic in sector 12 in Table 3 is half of the traffic of sector 12 in Table 2, and that the sum of the traffic in sectors 25 to 27 is equal to the traffic in sector 12 in Table 3. For many companies, this process of increasing sector traffic and adding new cell sites to offload congested sectors is typically performed manually, to allow for potential customization such as the introduction of a new wireless technology, changes in maximum spectral efficiency, deployment of small cells, changes in traffic growth rates, etc. If small cells are mandated to be deployed in a particular year, the network planner may assume that it will offload a certain percentage of the macro sector traffic. By analyzing OMs of small cells deployed in the field, the network planner might presume that a small cell will offload 15% of the traffic of a macro sector it was designed to relieve. Also, the capacity of the small cell will not exceed 15% of the maximum capacity of a macro sector (i.e. 1.635 Mbps in rural areas, 1.155 Mbps for urban areas), and an additional small cell is assumed to offload both existing small cells and the macro sector it is designed to relieve.

Perhaps the biggest complaint of this process is the tediousness of performing it manually. In the above example, planning the number of new cell sites for one year wasn't too arduous as only one new cell site was required. Performing this process for two or three years appears to be manageable, which is the time frame that engineers in most Engineering and Operations departments care about. The extension of this method for planning periods of up to 10 year is when this method becomes unwieldy as over 500 sectors will need to be added to meet the 40% per annum traffic growth forecast. For this reason, this exercise is typically performed once per year. The process is easy to teach and effective for its original purpose of determining which cell sites will reach congestion in the not too distant future (less than three years).

1.1 Historical Digression of Wireless Network Dimensioning

As wireless technology evolved, so has the method used to evaluate it. The first generation of commercial cellular service was exclusively designed to support voice telephony, where each conversation was allocated a dedicated channel occupying a certain amount of spectrum. Each MNO only had a limited number of channels. For example, in North America each MNO was allocated 25 MHz of spectrum (12.5 MHz for each of the downlink and uplink) supporting 416 channels, each occupying 30 kHz. Since the MNO needed to support thousands of simultaneous voice calls, each channel had to be used multiple times. A channel at one cell site will be used again at a cell site a certain distance away. Based on formulas that describe RF propagation, a plan was devised whereby each base station supported a fraction of the channels, typically one-seventh or 60 channels. Formulas that describe RF propagation showed that the seven-cell frequency reuse plan (N = 7) was sufficient to obtain a minimum *C/I* ratio[2] of 18 dB. Anyone familiar with analog FM technology will recognize that 18 dB incorporates

2 Carrier-to-interference ratio (*C/I*) is defined as the power of the desired signal over the sum of the power of all the interfering signals.

a huge margin of safety, since 4 dB is all that is required to eliminate static. A substantial safety margin is required to accommodate large variations in channel signal strength (up to 20 dB) when the mobile antenna is moved mere centimeters. This phenomenon is called Rayleigh or short-term fading, and a formula was derived that describes the probability distribution of the signal strength measurements. As commercial networks were deployed, it was discovered that the seven-cell frequency reuse scheme did not provide a sufficient margin of safety. Thus, sectored antennas were introduced at cell sites where each sector supported a third of the channels (20 channels). The engineers working in Operations and Engineering Planning had to calculate how much voice traffic each sector could support. It is too simplistic to assume that 20 channels can support 20 channels × 60 minutes/channel = 1,200 minutes of voice traffic for a given hour. If all 20 channels are in constant use, all new calls will be blocked. MNOs must specify the percentage of blocked calls that will be tolerated, and then use the appropriate formula to calculate the amount of traffic a sector can support.

The whole point of the previous paragraph is not to make you proficient in dimensioning first-generation cellular networks, but to demonstrate that engineering wireless networks did not need computer algorithms at one time. This changed when CDMA was first proposed. Before CDMA, the mobile equipment vendor could simply instruct the MNO to use a three-sectored seven-cell frequency reuse plan with 20 channels per sector and spare the MNO the trouble of researching wireless strategies. MNOs could make modifications to the basic strategy. For example, the engineers in Operations and Engineering Planning might observe that certain sectors carry more traffic than others and may expropriate channels from less-busy sectors to donate to overloaded ones. I'm not against customization, but it would be desirable to model its effects, so the benefits can be quantified.

When Qualcomm first proposed CDMA with its universal frequency reuse plan (N = 1), the industry realized it was a radical departure from the traditional frequency reuse plan employed by first-generation cellular networks. What got less attention was the effect CDMA had on the analysis and dimensioning of wireless networks. Monte Carlo simulations had to be used, as closed-form solutions were not possible for many calculations. Not wanting their customers to replicate their calculations, mobile equipment vendors provided stock recommendations that operators could use to dimension their networks. For example, a network vendor can simply state that one 1.25 MHz CDMA sector can support, on average, 25 Erlangs of voice traffic at 2% call blocking. Modeling more complicated CDMA networks was simpler than modeling first-generation analog networks, since the equipment vendor relieved the operator from the burden of consulting the Erlang-B table when calculating network voice-traffic capacity.

When MNOs deployed their CDMA networks, they realized busier sectors could support more voice traffic than the average sector capacity figure advertised by the manufacturers. This bit of good news was offset by the fact that the average amount of voice traffic per sector in the actual network was less than half the advertised sector capacity. While the advertised sector capacity was calculated based on simulations where each sector had the same number of subscribers, in actual networks the number of subscribers in each sector varied greatly. (For example, see Figure 1.) It was possible for the busier sectors to steal capacity from less busy sectors (much like in first-generation analog networks), but the overall network capacity was far less than what was advertised by the equipment manufacturers. This discrepancy led MNOs to question the claims of equipment vendors. MNOs placed less emphasis on manufacturers' advertised KPIs and started to rely on data collected from the network to formulate their own KPIs. Many engineers erred on the side of caution and interpreted the manufacturers' advertised average voice capacity figure as the maximum amount of voice traffic recommended for a sector. As a result, the realized average sector voice traffic was only a third of the manufacturers' claim.

As the Internet became prevalent, the wireless industry began to focus on supporting data applications. Instead of quoting the average amount of voice traffic a sector can support, equipment manufacturers began to quote the average amount of data traffic a sector can support, in bits/sec/Hz (bps/Hz). As with voice traffic, some sectors carried much more data traffic than others, and the realized spectral efficiencies of actual networks were far less than the spectral efficiency claimed by equipment manufacturers.

It is one thing to analyze traffic measurements from cell-site sectors to determine which ones are approaching congestion, it is quite another to develop a recommendation on the amount of traffic a sector can support based on a superficial analysis of those measurements. Noticing that busy rural sectors can carry more traffic than urban sectors, one may conclude some element about being in the country is responsible for higher spectral efficiency. There are no incentives for the network planner to determine what that is, but there is the hope that whatever caused it won't disappear as additional rural cell sites are added. If OMs indicate that rural cell sites have higher spectral efficiencies than urban ones, then it would be the responsibility of the skeptic to demonstrate why that fact should not be used for dimensioning the network.

There are potential pitfalls of naively analyzing network OMs when developing strategies to deploy new wireless technologies. In the case of small cells, early field trials may indicate that an ideally located small cell can offload up to 15% of the traffic of the macro sector it was designed to relieve. Not fully investigating the factors that may influence the amount of traffic offloaded, the network planner may be unable to find a reason to not use the figure obtained from the field trial for dimensioning. What is more deleterious are preconceived notions of how small cells should be deployed in a wireless network of macro cell sites. Intuition should not be relied upon, and it will be shown later that a more systematic method is indicated.

1.2 The Case for Automation

The process just outlined for dimensioning wireless networks is based on heuristics. A heuristic is any approach to problem solving, learning, or discovery that employs a practical method not guaranteed to be optimal or perfect, but sufficient for the immediate goals. I have many issues with the method, and I'll address one weakness: the cumbersome nature of its use when forecasting the number of new cell sites required over many years for a network already consisting of thousands of cell sites. It is common that traffic data for all cell sites are stored in a spreadsheet, and that this spreadsheet be modified for the purpose of dimensioning the growth of the network. If the planning horizon is on the order of a decade, it is likely that the number of cell sites will grow by more than an order of magnitude. It could take a team of engineers several months to update the spreadsheet to show the growth of the wireless network over the study period. Although I am not a fan of this method, at least I could automate the process, and save many person-years of engineering effort.

The MATLAB code fragment on the following page forecasts the growth of a wireless network over any number of years, using only sector traffic data from the initial year.

The program was executed twice: once with the rural sector traffic data in Table 1 as the starting traffic for year 1, and once with the urban sector traffic data. Annual traffic growth was assumed to be 40%. Figure 2 shows the percentage of sectors with a specified sector capacity utilization (congestion occurs at 100% utilization).

```
n = 10; %number of years in the study period
incr = 0.4; %annual increase in traffic
offload = 0.5;  %amount traffic offloaded from the congested sector
loadsoy = zeros(1000,n+1); %sector traffic at the start of the year
loadeoy = zeros(1000,n);   %sector traffic at the end of the year
%sector traffic at the start of the study period. From Table 1.
year1 = [4.1 4.4 1.8 1.5 2.2 7.6 1.1 5.4 2.0 2.1 1.9 10.9 2.4 2.4 1.6 3.2
1.6 4.7 5.9 1.6 1.5 3.3 1.0 1.7]; %year 1 rural cluster sector traffic
%year1 = [4.0 4.7 2.7 4.6 4.4 3.3 4.1 4.2 4.6 4.6 3.1 3.1 3.8 3.7 3.8 5.0
2.9 4.6 2.9 7.7 5.4 3.6 4.2 2.3]; %year 1 urban cluster sector traffic
capacity = max(year1);  %state the maximum capacity of the sector
loadsoy(1:length(year1),1) = year1'./capacity;
m = length(find(loadsoy(:,1) > 0));
rng('default');
for k = 1:n
    loadeoy(:,k) = loadsoy(:,k)*(1+incr);
    loadsoy(:,k+1) = loadeoy(:,k);
    exhaust = find(loadeoy(:,k) >= 1);  %find sectors that are congested
    for kk = 1:length(exhaust)
        m = m+1;
        offloadtemp = rand(1,2);
        offloadsort = sort(offloadtemp);
        offload1 = offloadsort(1);
        loadsoy(m,k+1) = loadsoy(exhaust(kk),k+1)*offload*offload1;
        m = m+1;
        offload2 = offloadsort(2)-offloadsort(1);
        loadsoy(m,k+1) = loadsoy(exhaust(kk),k+1)*offload*offload2;
        m = m+1;
        offload3 = 1 - offloadsort(2);
        loadsoy(m,k+1) = loadsoy(exhaust(kk),k+1)*offload*offload3;
        loadsoy(exhaust(kk),k+1) = loadsoy(exhaust(kk),k+1)*(1-offload);
    end
end
```

Although the shapes of the histogram for rural and urban traffic in year 1 are very distinct (see Figure 1), after 10 years of traffic growth, the shapes of the histograms for rural and urban sector capacity utilization converge. The average sector capacity utilization is 37% for both rural and urban. At the start of the study period, the average rural sector capacity utilization was at 29% while urban was 53%. It appears that the average sector capacity utilization at the start of the study period has no effect on the average sector capacity utilization at the end. People who might consider using this method for dimensioning network growth should be aware of this detail.

Since the histograms for rural and urban sector capacity utilization are similar, the ratio of the mean traffic to the sector capacity should be the same for both morphologies. Since the capacity for rural sectors is higher than for urban sectors, the average traffic for rural sectors will also be higher than for urban sectors (4 Mbps for rural sectors, 2.83 Mbps for urban). Note from Table 1, the average traffic for rural sectors was initially lower than for urban sectors (3.2 Mbps for rural sectors, 4 Mbps for urban).

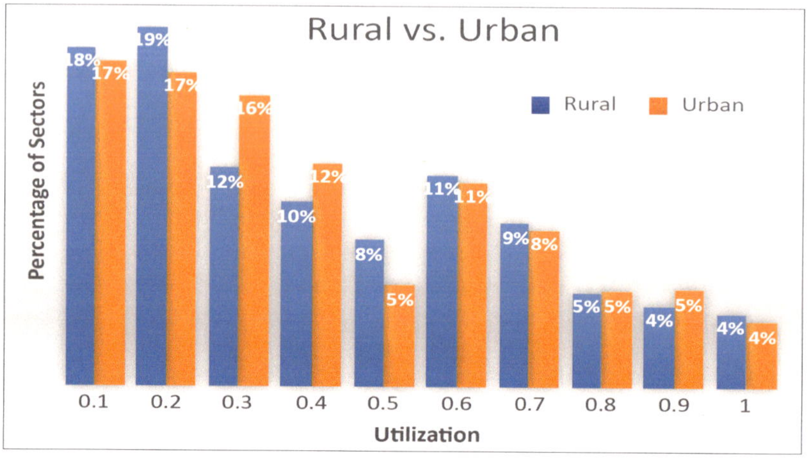

Figure 2: *Percentage of sectors within a specified utilization after 10 years of traffic growth.*

By using a computer algorithm to address the labor intensiveness of the heuristic method, I've inadvertently uncovered at least two issues:

- the histogram of the sector traffic converges to a peculiar shape that is independent of the current traffic distribution
- the spectral efficiency of rural cell sites increases over time while the spectral efficiency of urban cell sites decreases.

To reiterate, these problems do not appear when the heuristic method is restricted to its original role of identifying cell sites that will potentially experience traffic congestion over a short time frame (less than three years).

1.3 An Alternative Dimensioning Method

The biggest problem with relying exclusively on OMs to dimension wireless networks as described in the above method is that some of the assumptions are treated as axioms.[3] In the remainder of the book, I will describe an alternative method to dimension wireless networks.

3 An axiom or postulate is a statement that is taken to be true, to serve as a premise or starting point for further reasoning and arguments.

It has a different set of axioms, ones based on a less-questionable set of assumptions. I will also explain the theory used in developing this alternative method.

To make the example more concrete, I will use the proposed model to value spectrum for a particular scenario. The value of a spectrum licence is typically measured in $/MHz/POP (dollars by megahertz by population). The greater the amount of spectrum in terms of bandwidth (measured in MHz) and the greater the number of people the spectrum licence covers (measured in POP), the more valuable the spectrum licence becomes. For example, if spectrum costing $10 million occupies 10 MHz of bandwidth and covers 1 million people, then the normalized spectrum value is $10 million/10 MHz/1 million people = $1/MHz/POP. Conversely, if the value of the spectrum is $1/MHz/POP, 10 MHz of spectrum covering 1 million people is worth $10 million, twice as much as 10 MHz of spectrum covering 500,000 people or 5 MHz of spectrum covering 1 million people. Spectrum valuation is most often associated with spectrum auctions and the acquisition of spectrum licences.

Assume for this analysis that the MNO has deployed a cellular network in an area where it has existing spectrum licences and all frequency bands being considered reside in the same spectrum band.[4] One method used to find the value of a new spectrum licence is to measure how much capital expenditure (CAPEX) it could save if the company acquired the additional spectrum. An MNO needs to increase network capacity due to the A) increasing number of subscribers and B) increasing usage per subscriber. As mentioned, increasing capacity can be achieved by either increasing the number of cell sites or configuring radios to utilize unused spectrum (if unused spectrum is available). In this particular scenario, it is cheaper to configure radios to utilize unused spectrum than to build a new cell site.[5] Therefore, one could value spectrum by how many cell sites are not built due to the addition of spectrum. Note that the heuristic method based on OMs could be used to value spectrum: one determines the amount of network assets required to support a given level of traffic with the existing inventory of spectrum and then recalculates the amount of network assets when the additional spectrum is acquired. The amount of network assets required is directly related of the amount of traffic the network is forecasted to carry, and the CAPEX avoidance from the addition of new spectrum would also increase with higher traffic forecasts. Hence, it is possible that the person performing spectrum valuation would run the exercise with a range of forecasted traffic growth. This will make the

4 The assumption that all spectrum reside in one band is critical in this analysis, as we will see later how the number of spectrum bands can affect results.
5 Later in this book, we will encounter a scenario where this is not the case.

people using the heuristic method very busy, since there are possibly tens of thousands of cell sites that need to be analyzed.

Here is the alternative and—I will show—arguably simpler method of performing the same analysis.

The company's strategic wireless planner will tell you the average number of gigabytes per month (GB/month) each MHz of deployed spectrum can support. This is called the spectral efficiency of the network, although most people are more familiar with spectral efficiency quoted in bit/sec/Hz[6]. For purposes of concreteness, let's say that air-interface technology X can support on average 30 GB/month for each MHz of spectrum the radio is configured to use. Assume that the mobile network operator has 50 MHz of downlink spectrum and 50 MHz of uplink spectrum. A typical macro-cellular cell site typically supports three sectors, which means 150 MHz of spectrum can be configured for downlink transmission (which is typically the busier link). The cell site can support 150 MHz × 30 GB/month/MHz = 4,500 GB/month of traffic. Now assume that 2 × 10 MHz (10 MHz down link and 10 MHz uplink) is being offered in the same spectrum band currently used to support wireless service. If the mobile network operator acquires the additional spectrum, it will be able to configure the radios to support 180 MHz of spectrum on each cell site, thus supporting 180 MHz × 30 GB/month/MHz = 5,400 GB/month of traffic, or 900 GB/month more traffic than when only 50 MHz was available.

Once again, it is easy to see that the number of cell sites saved is a function of the total amount of traffic the network is forecasted to carry. If the network is forecasted to carry 900,000 GB/month, then 200 cell sites are required when only 50 MHz of downlink spectrum is available, and 167 cell sites when 60 MHz is available. If Operations claims that a cell site costs $600,000, then the CAPEX savings of not deploying 33 cell sites is $20 million.[7] This analysis takes little time and could, in a real-world scenario, impact capital investment decisions on the order of tens of millions of dollars.

More generally, the network with only 50 MHz of spectrum requires 20% more cell sites than the network with 60 MHz of spectrum. Or, to increase capacity by 20%, you can either increase the amount of spectrum by 20% or increase the number of cell sites by 20%.

The calculation is so simple, you can nearly do it in your head, and that basic calculation is the foundation of all the business and economic value

6 It will be shown later how to convert bit/sec/Hz to GB/month/MHz and vice versa.

7 The mobile network operator will also realize OPEX savings from not building 33 cell sites.

that this method delivers.

Next, suppose the R&D engineer tells you that the spectral efficiency of 30 GB/month/MHz is applicable if the wireless network is strictly co-channel interference limited, i.e. co-channel interference is much stronger than thermal noise, a situation found in urban scenarios. (Don't worry if these terms do not make sense, as they will be explained later.) If the network is not strictly co-channel interference limited, the engineer tells you that the spectral efficiency will be lower (25 GB/month/MHz, for example). Or perhaps the 30 GB/month/MHz spectral efficiency for strictly co-channel interference limited scenarios was much too optimistic and the recommendation is to lower it to 25 GB/month/MHz. In any case, the recalculation using the new spectral efficiency number is fairly simple.

The claim that the capacity of a wireless network can be simply calculated by adding the spectrum bandwidth of the deployed radios and multiplying that by the average spectral efficiency of the radio (30 GB/month/MHz in our example) is overly simplistic. Determining the total spectrum bandwidth of all the deployed radios is trivial, yet determining the spectral efficiency of the radios is another matter.

Before getting into the details of all the factors that affect the average spectral efficiency of the radios, I suggest a simple thought experiment that supports the alternative simple network capacity calculation. If you double the total spectrum bandwidth of the deployed radios, does the spectral efficiency of the radio change? Perhaps you can think of situations where it does but, in most scenarios, it is likely that the spectral efficiency of the radio doesn't change and all you have done is double the capacity of the network. Continuing with this thought experiment, the Operations department tells you the wireless network is at capacity. You also know the total spectrum bandwidth of the deployed radios in the network. You can just divide the monthly network traffic (in GB/month) by the spectrum bandwidth of the deployed radios (in MHz) to get the spectral efficiency of the deployed assets (in GB/Month/MHz).

The one big difference between the two dimensioning methods is the choice of parameters. The first method, based on OMs, uses the maximum spectral efficiency of a cell-site sector (1.09 bps/Hz for rural sectors, 0.77 bps/Hz for urban sectors) while the alternative method uses the average spectral efficiency of a cell-site sector when the network is at maximum capacity (30 GB/month/MHz or, as it will be shown later, 0.3 bps/Hz). Another difference is that the heuristic method requires a specific set of data in the amount of sector traffic before the process of dimensioning can begin. The alternative method is somewhat more abstract, where analysis can proceed without specifying sector traffic.

1.4 Exploratory Data Analysis

> *Exploratory data analysis (EDA) seeks to reveal structure or descriptions, in data. We look at numbers or graphs and try to find patterns. We pursue leads suggested by background information, imaginations, patterns perceived and experience with other data analyses.*
>
> — *Persi Dianconis*[8]

Referring to Table 1, the traffic for each sector can be considered a sample from a population of all values of sector traffic. When traffic data is viewed in this manner, statistical inference can be used to generate possible insights and help with decision making. John Tukey, whom we met in the Introduction, developed a statistical tool called exploratory data analysis (EDA) while working at AT&T Bell Laboratories. It is one of the most prominent tools today. Tukey was concerned that statistics was too narrowly focused on hypothesis testing as envisaged by Ronald Fisher. According to Tukey, raw data should be analyzed first before the development of a candidate hypothesis. The histogram (see Figure 1 for example) is one of the tools of EDA, along with the analysis of central tendency (mean or median), spread (range, variance and percentile) and shape (skewness).

One technique that I often used was to fit a probability distribution to the observed data traffic histogram such as the one shown in Figure 1. There were many distributions that can be used to model the dispersal of sector data traffic such as log-normal, gamma, truncated normal or Weibull distributions. Unfortunately, I could not find a mechanism to explain the prevalence of those distributions. Had I found a mechanism, I wonder if the discovery would have been of little more than academic interest. The technique that I will develop in subsequent chapters assumes that any unknowns such as the geographic distribution of traffic can be treated as an adjustable parameter to be modified so its effect on KPIs can be observed.

One benefit of employing statistics is the confirmation of using the average sector data traffic in a cluster operating at capacity and not the amount of traffic in the busiest sector for the purposes of network dimensioning.[9] Using statistical nomenclature to facilitate the communication of ideas among various stakeholders is another benefit. However, there were many questions that statistics could not answer, such as the capacity gains that can be achieved by altering antenna

8 From *Data Tables, Trends, and Shapes*, edited by D. C. Hoaglin, F. Mosteller and J. Tukey, John Wiley, New York, 1985.

9 This assertion will be examined later when the stochastic method is explained.

patterns, increasing base-station transmit power, decreasing inter-site distance, or increasing receiver sensitivity. Those questions will have to wait for the development of the alternative method that I will outline in the next several chapters. Statistics will not be totally excluded, as it will be used in a supporting role by judging the plausibleness of a proposed planning technique. I have already commented on how the heuristic method described at the start of this chapter produces a histogram with the same unique shape regardless of the current distribution of the sector traffic. (See Figure 2). Also, the histogram appears to be bimodal,[10] which is something I have not observed in actual networks. Perhaps the heuristic method could be modified such that the distribution of sector traffic does not change when new cells are added, or evolves in a manner that produces histograms resembling those found in actual networks.

10 A bimodal histogram has two peaks.

2 | Necessary Preliminaries

There are countless textbooks that teach the theory of communications. Most of these assume you have a good understanding of calculus, linear algebra, probability and other topics in mathematics that you have probably forgotten or never taken. Coincidentally, machine learning also requires those same math skills, plus the ability to program. I won't be teaching machine learning, and the communication concepts I describe only require high-school level mathematics. This book is not a comprehensive guide to the mathematics of network planning, but rather it covers the necessary material needed to understand the theory behind the stochastic method described in the next chapter.

2.1 Elements of Information Theory

Any method intended to assist in the dimensioning of wireless networks or to guide wireless strategy development should include in its foundation the relationship between signal-to-noise ratio and capacity. This relationship can be expressed in an equation or in the form of a graph. If an equation is chosen, the most important equation in information theory is called the Shannon-Hartley equation:

$$(1)\quad Capacity/Channel = Bandwidth * \log_2\left(1 + \frac{Signal\ Power}{Noise\ Power}\right). \quad [bit/sec]$$

The equation looks simple, but its derivation is non-trivial. You don't lose much by not knowing the proof. Just use it and trust that it works. The equation tells us the highest possible bit rate that can be transmitted across a channel for a given bandwidth and signal-to-noise ratio. Note: if we divide both sides by the bandwidth, we get the spectral efficiency of the channel.

$$(2)\quad Capacity/Channel/Bandwidth = \log_2\left(1 + \frac{Signal\ Power}{Noise\ Power}\right). \quad [bit/sec/Hz]$$

With the advent of turbo and LDPC (low-density parity-check) codes, it is possible to send data at a rate approaching the Shannon limit. Note how the Shannon-Hartley equation tells you how to increase the data rate of the channel. You can:

- Increase the signal-to-noise ratio
- Increase the spectrum bandwidth
- Increase the number of channels.

Claude Shannon introduced the above equations in his seminal 1948 paper "A Mathematical Theory of Communications." Communications engineers are keenly aware of the effects of signal power and noise on the data rate of a communication channel, and had developed practical techniques to increase spectral efficiency even before Shannon published his equation. Most engineers will find the derivation of the Shannon-Hartley equation very abstract as it gives little in the way of practical advice on how channel capacity can be achieved. However, without the capacity equation, these engineers would have been working in a fog, not knowing how much room there is for improvement. Although Shannon derived the capacity equation using an analytical approach, engineers used trial and error to invent numerous modulation and coding schemes that were tested against the capacity equation using computer simulations.

We have previously used an alternative measure for spectral efficiency: GB/month/MHz. In performing this conversion from bit/sec/Hz to GB/month/MHz, we need to understand how traffic varies during the month and the concept of busy hour. Data traffic is not constant, but varies significantly during the day. (See Figure 3 for an example of a week's worth of data traffic for a particular cell sector.) There are different definitions for the term "busy-hour traffic" but suffice it to say it is a measure of the amount of traffic in the busiest hour of the day. If data traffic were evenly distributed throughout the day, the monthly data traffic would be 720 times (30 days times 24 hours per day) greater than the busy-hour traffic. Due to the unevenness of traffic throughout the day, it turns out that the monthly data traffic for a sector is typically 240 times the busy-hour traffic. That means if one is given the monthly data traffic for a sector, the busy-hour traffic can be calculated by dividing the monthly traffic by 240. Ideally, it would be beneficial to have the ratio of monthly traffic to busy-hour traffic be as big as possible, since the network must be designed to carry the busy-hour traffic.

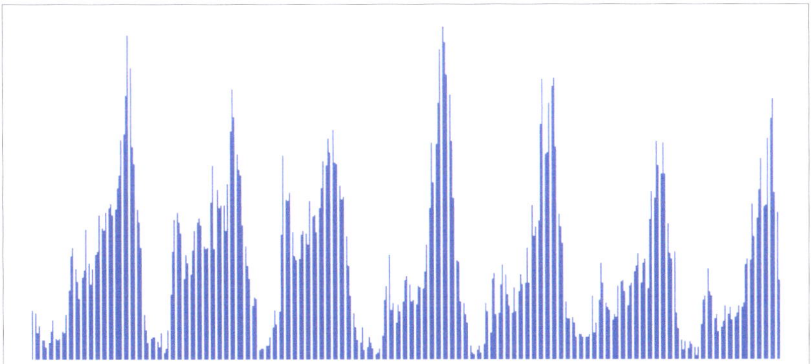

Figure 3: *Weekly data traffic for a wireless cell sector*

To do the conversion from bit/sec/Hz to GB/month/MHz, we first note that 1 bit/sec/Hz is the same as 1 Mbps/MHz. To get GB/hour, take 1,000,000 (i.e. 1 Mb) and divide it by 8 to get bytes/sec, then multiply by the number of seconds in an hour (3,600) to get bytes/hr and then divide by 1024^3 to get GB/hr.[11] Hence 1 Mb/sec is the same as 0.419 GB/hr. Since monthly traffic is approximately 240 times the busy-hour traffic, 1 Mb/sec during the busy hour is $0.419 \times 240 = 100.6$ GB/month/MHz or 100 GB/month/MHz, since the 240 monthly busy-hour figure was just an approximation.

From Chapter 1, the hypothetical R&D engineer told the client to use 30 GB/month/MHz as the spectral efficiency for a particular air-interface standard. This is equivalent to 0.3 bit/sec/Hz. Or perhaps the thinking was that the spectral efficiency is 0.25 bit/sec/Hz, but the number of busy hours per month is 286.33. This will also give 30 GB/month/MHz. Unfortunately, this kind of ambiguity happens all the time. The number of busy hours in a month is one of most variable parameters, due to:

- The number of subscribers in the sector. The more subscribers in the sector, the lower the variation in hourly traffic. When hourly traffic is more uniform, the number of busy hours in the month increases. Ideally, we would like every hour in a month to be busy hours, as that would indicate every hour has the same amount of traffic.
- Differing definitions. Some people use a network number and, since a network consists of many sectors, the number of subscribers in the network is far larger than the number of subscribers in a sector. The greater the number of subscribers, the lower the variation in hourly traffic, the greater the number of busy hours per month. See the previous point.
- There is no reason why the busy hour number cannot be any particular value between 1 and 720 (or 744 for a month with 31 days). If there is only one busy hour in the month, all the monthly traffic occurs during that one hour. If there are 720 busy hours in the month, each hour of the month carries the same amount of traffic. All numbers between 1 and 720 are completely valid. The only way to get a value for the monthly number of busy hours is to perform field measurements. And there is no reason for that number not to change. The best you can do is to either take an average of a collection of

11 It still drives me nuts that $1GB = 1024^3 = 1{,}073{,}741{,}824$ bytes and not 1,000,000,000 bytes. All major telecommunication providers use the binary definition of gigabyte and not the SI definition. The only good thing about using the binary version is that consumers get more bytes for their gigabytes.

measurements or, if you are more conservative, take the 5th or 10th percentile[12].

In summary, don't fret too much about the busy-hour number. Perhaps it is best to state the number of busy hours in a month as an assumption and show how network capacity and spectral efficiency changes when the busy-hour assumption changes.

Instead of using the Shannon-Hartley equation, we could use charts where data-link speeds are specified with respect to signal-to-noise ratios. You use graphs instead of the equation because they include inefficiencies introduced by various factors, such as short-term fading, the effects of doppler, control channel impact, the effects of transmitting pilot signals, etc. Below is a chart generated by field test measurements performed by an equipment vendor. (See Figure 4.) It includes graphs for different MIMO antenna configurations. For those not familiar with MIMO (Multiple Input Multiple Output), it is a fairly recent development in which multiple antennas are placed at the transmitter and receiver to increase the capacity of the wireless link. Like many other areas of wireless communication technology, one could spend years studying and investigating MIMO technology. Or one could trust that it works and use the graphs to determine network capacity.

Figure 4: *LTE data speeds versus signal-to-interference-to-noise ratio*[13]

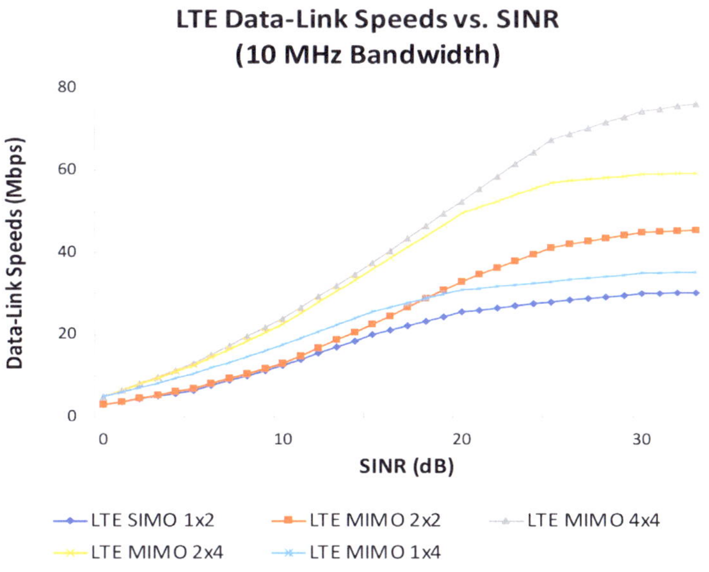

12 The 5th percentile is the level at which the lowest 5% of measurements fall below. The lower the monthly busy hour number, the less monthly traffic that can be carried.

13 From "HSPA to LTE-Advanced: 3GPP Broadband Evolution to IMTAdvanced (4G)", Peter Rysavy, *Rysavy Research*, September 2009.

2.2 Resource Utilization

All data-networking equipment has a finite amount of resources that can be used to carry data. For example, T1 has 24 time slots in every 125 ms frame, each carrying a separate digitized PCM voice call. The original IS-95 CDMA (Code Division Multiple Access) air-interface standard has 64 orthogonal Walsh Codes initially designed to each carry a separate digitally encoded voice call. Resource utilization or utilization refers to the percentage of resources (such as time slots) carrying data.

Although the method I've developed works with any digital air-interface, I will primarily use LTE to avoid dealing with too many generalities. In LTE, data is transported in Resource Blocks which consist of 12 consecutive subcarriers lasting seven consecutive OFDM symbols (0.5ms or one slot). Hence, a Resource Block consists of $12 \times 7 = 84$ Resource Elements, the majority of which carry data. (See Figure 5.) Since each subcarrier occupies 15 kHz of spectrum, a Resource Block occupies 180 kHz. In a 20 MHz LTE carrier, there are 100 Resource Blocks occupying 18 MHz of spectrum (there are two 1 MHz guard bands on each side of the LTE carrier). Similarly, there are 50 Resource Blocks in a 10 MHz LTE carrier and 25 Resource Blocks in a 5 MHz LTE carrier.

Figure 5: *LTE downlink frame structure.*

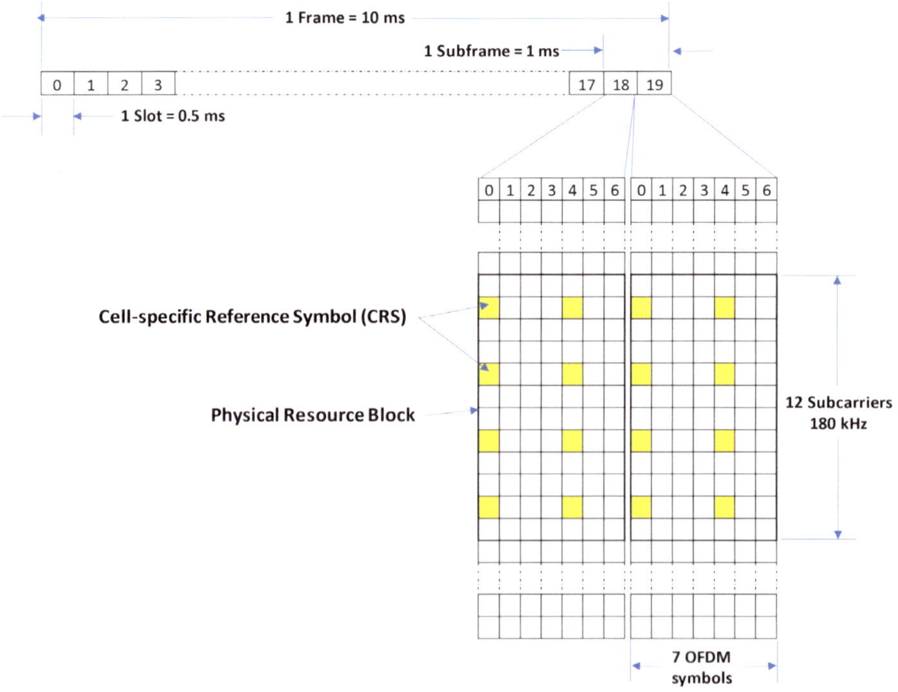

The capacity of a wireless network is the maximum amount of traffic it can support. Intuition would suggest that a network will carry the maximum amount of traffic when all Resource Blocks are carrying data (i.e. 100% Resource Block utilization), and that would be correct most of the time. At 100% Resource Block utilization, co-channel interference will be at its maximum and the amount of data each Resource Block can carrying will be less than when there is less traffic and less co-channel interference. In most scenarios, it pays to fill each Resource Block with data if maximizing network traffic is the goal. However, it is instructive to examine cases where this is not optimal. Regardless, no operator would operate a network in this manner, because the user-perceived throughput would be 0 Mbps. (See section 2.3.)

Another interesting aspect of the LTE Resource Block is approximately 10% of the Resource Elements (eight out of 84) are used for Cell-specific Reference Signals (CRS). Cell-specific Reference Signals are used for:

- cell search and initial acquisition
- downlink channel estimation for coherent demodulation/ detection at the UE
- downlink channel quality measurements.

When there is no data to transmit, the LTE radio is still transmitting CRSs, creating co-channel interference in other sectors. The amount of interference created by an empty Resource Block is approximately 10 dB less than a Resource Block carrying data.[14]

A possible reason for the heavy use of CRS in LTE is because only full-buffer traffic models were considered when developing LTE. If capacity was the main focus of 3GPP, then the most important scenarios are the ones where 100% of the Resource Blocks are carrying data. The impact on UPT (user-perceived throughput) of transmitting the CRS in idle frames when there was less than 100% PRB (Physical Resource Block) utilization was not given much consideration. In 5G, the CRS has been eliminated, thus reducing co-channel interference from idle Resource Blocks, dramatically improving user-perceived throughput. The impact of CRS will be investigated more thoroughly when we analyze various network strategies.

2.3 Capacity, Traffic and User-Perceived Throughput

This is one of the more important sections in this book. The topics here aren't very difficult, so it is surprising how little attention they generate in network company offices. Mobile network operators are chiefly concerned about how much traffic the wireless network can carry. They

14 L. Liu, Y. Li, B. L. Ng and Z. Pi, "Radio Resource and Interference Management for Heterogeneous Networks," *Heterogeneous Cellular Networks*, John Wiley & Sons Ltd, 2013.

are also concerned about the user-perceived throughput (UPT), which is how fast the network appears to the subscriber downloading data. As you will see, there is a trade-off between how much traffic the network carries and the UPT experienced by subscribers.

Historically, most of the emphasis was on how much traffic the wireless network can carry. This meant that every macro sector radio is transmitting data all the time, or, in industry parlance, the buffer or queue is always full. As will be shown, in actual networks the UPT is zero whenever the buffer is full: not a satisfactory situation for the subscriber. I don't blame the industry for modelling this scenario, since mobile network operators are always looking for higher capacity. However, some capacity-enhancing features only work when the input buffer is practically full. Vendors devise proprietary scheduling algorithms that can increase the capacity of a radio link under these conditions and market these features to mobile network operators. The problem is that oft-quoted spectral efficiencies are never met in actual network deployments, thus a different spectral efficiency number needs to be derived.

2.3.1 Harmonic Mean

Before talking about the capacity of a cell-site sector, we need to discuss a type of mean that isn't mentioned much in statistic books: the harmonic mean. I'll start with a simple illustration.

> **Example:** You are a mobile subscriber driving around the city and downloading a bunch of files. In the first location, you download a 1.2 Gb file (1.2×10^9 bits) at 20 Mbps. You drive to a second location and download another 1.2 Gb file at 60 Mbps. What is the average download speed for downloading 2.4 Gb?
>
> **Solution:** Hopefully you didn't add 20 and 60 and divided by 2 to get 40 Mbps. That would be the right answer if you spent the same amount of time downloading at 20 Mbps and at 60 Mbps. The solution is to find how long it takes to download 2.4 Gb. It takes 1,200 / 20 = 60 seconds to download 1.2 Gb at 20 Mbps. It takes 1,200 / 60 = 20 seconds to download 1.2 Gb at 60 Mbps. Therefore, it takes 60 + 20 = 80 seconds to download 2.4 Gb. The average download speed is (1200 + 1200) / 80 = 30 Mbps.

Note: we could have selected a different file size and still have gotten the same harmonic mean of 30 Mbps if the size of the file at both locations is the same. This broaches another topic related to the users' behavior under different link-speed conditions. Do people download more when data-link speeds are faster, or is user behavior independent of data-link speeds? From a mobile network operator point of view, you would like the subscriber to download more data when the data-link speeds are faster, because it would consume less resources. Some scheduler implementations allocate more time to subscribers whose RF link conditions are improving. This will theoretically improve overall link capacity. It will be discussed later that, in practice, the gain from these advanced scheduling algorithms may not be as much as first envisioned. The conservative approach is to assume that subscribers download the same amount of data independent of data-link speeds. This approach also reflects reality in most cases, unless the mobile network operator manages a constantly congested network.

There is another reason to assume that each subscriber downloads the same amount of data regardless of data-link speeds. When we develop the model, cell sites that carry more data can be simulated by assigning more subscribers to that cell site. A cell site that carries twice as much traffic as the average cell site will have twice as many subscribers associated with it.

The general definition for the harmonic mean is:

$$h = \frac{n}{\frac{1}{s_1} + \frac{1}{s_2} + \cdots + \frac{1}{s_n}}. \tag{3}$$

Where s_i is the speed of the data link. The proof is pretty simple. Invert the above equation to determine the average amount of time to transmit 1 bit.

$$\frac{1}{h} = \frac{\frac{1}{s_1} + \frac{1}{s_2} + \cdots + \frac{1}{s_n}}{n}. \tag{4}$$

If the average time to transmit 1 bit is, for example, $1/h = 1 \times 10^{-6}$ seconds, then the data link can transmit $h = 1 \times 10^{6}$ bits/sec.

I'm not going to show the proof, which can be found from many different sources, but the harmonic mean is always less than the arithmetic mean.

Let's now analyze the capacity of a cell-site sector. The focus will be on the downlink since it accounts for the vast majority of traffic, although the analysis can be extended for the uplink. The typical cell-site sector covers a large area, and the possible signal-to-noise ratio will vary vastly depending on the location of the point within the sector's coverage area.

In addition, each location will see its signal-to-noise ratio change as a function of time, as the source of co-channel interference (neighboring sectors transmitting signals to other mobiles on the same frequency band) varies depending on whether these other radios have data to transmit. As the signal-to-noise ratio changes, so does the speed of the data link. (See the Shannon-Hartley equation.) In determining the data-link speed of a particular location, we can take a sample of the speed of the link at different points in time. For example, for one second the link to a particular location can support 60 Mbps, then in the next second the link can only support 20 Mbps because a radio in the adjacent sector starts to transmit data to another mobile, causing co-channel interference. In this scenario, we do not use the harmonic mean since the amount of data being downloaded in the first second is not the same as the amount of data transmitted in the next. In this situation, we use the simple arithmetic mean. A total of 80 Mb was transmitted in two seconds, hence, the average data speed '*s*' is (20 + 60) / 2 = 40 Mbps. The average data-link speed is estimated for every subscriber location in the sector, and then we use the following equation to determine the capacity of the sector:[15]

(5)
$$\rho_c(X) = \frac{|V(X)|}{\int_{V(X)} 1/s\big(SINR(y)\big)dy}.$$

Or instead, you can randomly select a set of subscriber locations in the sector and take the harmonic mean of the average data-link speeds s_i for those points. That is:

(6)
$$\rho_c = \frac{n}{\frac{1}{s_1} + \frac{1}{s_2} + \cdots + \frac{1}{s_n}}.$$

The amount of traffic in a sector depends on the percentage of the radio resource utilized, i.e. its utilization θ:

(7)
$$\rho = \rho_c \theta.$$

Equations (6) and (7) are probably the two most important in this book. If these equations were simple to calculate, then network dimensioning and developing wireless strategy would be easy. Unfortunately, calculating ρ_c and ρ can be extremely challenging, since the data-link speed at a specific instance and location is affected by co-channel interference from other sectors and can be determined only by downloading data, which will in turn cause co-channel interference in the other sectors where different locations are also downloading

15 This is the formal definition of $\rho_c(X)$. No reference will be made to this equation in the remainder of this book.

data. Hence, ρ_c is a function of the amount of traffic in the wireless network. The greater the amount of traffic, the greater the amount of interference, the lower the capacity.

There is much confusion between the following two concepts: first, sector (or network) traffic as just defined and, second, user-perceived throughput, which is a measure of how fast the network appears to the subscriber. The confusion may arise because both traffic and user-perceived throughput are measured with the same units, bit/sec. Traffic is a measure of how much data is sent across a link, while UPT is an indication of how fast the network feels. Although traffic and UPT are different, they are related. The more traffic a sector carries, the lower the UPT for that sector. Mobile network operators typically want as much traffic traversing the network as possible (unless the subscriber base has signed-up for flat-rate pricing), while users desire faster user-perceived data speeds. Unfortunately, you can't have both.

2.3.2 Little's Law

I stated in the previous section that mobile network operators must make a trade-off between traffic and user-perceived throughput. To see how the two concepts are related, we need to refer to Little's Law, which is part of the foundation of the method I've developed:

$$L = \lambda W. \tag{8}$$

Where L is the average number of packets in the system, λ is the average arrival rate into the system and W is the average time the packet spends in the system. Little's Law is one of the fundamental equations in the field of queueing theory. I'm not going to teach queueing theory, because that would require a fairly large book on its own. However, most people have a general idea of what a queue is: go to a fast-food drive-through during breakfast or lunch and you will understand that the busier the drive-through, the longer the wait times. The same principle applies to the input buffer of a radio. Instead of cars, data packets line up to be transmitted from the radio to the subscriber; each must wait its turn in the queue or buffer. The longer the queue, the slower the link will appear to the subscriber. The total time the data packet spends in the system consists of time in the queue and the time being actively transmitted over the air. The average data rate perceived by the subscriber can be calculated by dividing the average data packet size by the average time the data packet spends in the system (queue plus transmission time). For example, assume that the average data packet size is 1,000 bits and it spends on average

1 millisecond in the system, then the user-perceived throughput is simply 1000/.001 = 1,000,000 bit/sec or 1 Mbps.

Unlike the Shannon-Hartley equation, Little's Law is very easy to prove. When Little first proposed his formula, it was stated without proof. Subsequently, Little and other researchers found a proof for his formula, but it involved some complicated mathematics. However, I found a simple proof in Sheldon Ross' probability book *Introduction to Probability Models*, although I'm not certain if he is the original author of the proof. It requires no calculus or knowledge of stochastic processes, much like this book.

Imagine you are a consultant. An operator of a fairground hires you to estimate the revenue the fairground is expected to generate. You think about the problem and come up with a solution. You determine that the average family spends $1 for every minute at the fairground. You need to estimate the average number of families (L in Little's Law) at the fairground and multiply it by $1/minute to determine how much money the fairground is generating every minute. You present your findings to the fairground operator, but he is slightly confused. Unbeknownst to you, he has hired a second consultant to work on the same problem. The other consultant found a different solution. He too found out that the average family spends $1 per minute at the fairground, but his method differs: instead of estimating the average number of families at the fairground, he estimates the length of the average visit in minutes (W in Little's Law) and multiplies it by the rate at which the average family spends ($1 per minute) to determine the spend for the day. He then finds out the rate at which families arrive at the fairground ('λ' in Little's Law). He multiplies the arrival rate (families per minute) and the average revenue per family to determine how much money the fairground is generating per minute. You and the fairground operator agree that both answers appear to be correct. Hence, it appears that $L = \lambda W$. The nice thing about this proof is that it didn't involve calculus or any mention of stochastic processes. The fairground analogy would never be accepted as a proof among mathematicians, but in the business world, who cares? Later in this book, I'll provide a method of calculating Equations (6) and (7) that requires no understanding of stochastic calculus.

Now that we have Little's Law, we will use it to derive the relationship between traffic and user-perceived throughput.

2.3.3 Derivation of User-Perceived Throughput

Let $1/\mu$ represent the average size of the data packet to be transmitted, typically measured in bits. As previously mentioned, user-perceived throughput (denoted by r) is defined as the average data packet size divided by the average time it spends in the system (time spent in the queue plus time being transmitted), or $r = (1/\mu)/W$. From Little's Law, we know that $L = \lambda W$. Thus, $r = (\lambda/\mu)/L$. λ is the rate at which data packets of average size $(1/\mu)$ are arriving in the system, and (λ/μ) is the average date rate of the traffic (measured in bit/sec) for which we've already defined a symbol, ρ. For example, if λ is 500 packets per second and $1/\mu$ is 1,000 bits, then the traffic $\rho = (\lambda/\mu) = 500 * 1{,}000 = 500{,}000$ bit/sec. So, the equation for user-perceived throughput — $r = (\lambda/\mu)/L$ — is just the traffic divided by the average number of packets in the system. Note that it is not the capacity of the link divided by the average number of users in the sector[16], but the traffic divided by the average number of packets in the system. Now we need to know the average number of packets in the system. We get this by using the link utilization, the percentage of the link that is actually carrying traffic (denoted by θ). If the link is fully utilized, then $\theta = 1$, else $\theta < 1$. Sector utilization is one of the operational measurements provided by the wireless network.

Now another simple formula, for which I will not provide the proof. If one assumes a Poisson point process (so the arrival of data packets is completely random), then one can find the average number of packets in the system using this simple formula:

$$L = \frac{\theta}{1-\theta}. \qquad (9)$$

The Poisson point process is also part of the foundation of the method I've developed. Finishing off the derivation, $r = \rho / L = \rho\,(1 - \theta) / \theta = \rho_c - \rho$. The user-perceived throughput is just the capacity of the system minus the actual carried traffic.

$$r = \rho_c - \rho \qquad (10)$$

As simple as this expression is, it was not widely used before the widespread adoption of the Internet.[17] Academics developed a sophisticated field of study called queueing theory decades before the Internet, but they used other metrics to characterize network congestion, such as increased service times. Note the trade-off between traffic and UPT.

16 In many simulations, average user throughput is defined as the capacity of the link divided by the average number of users in the sector. It is assumed that each user in the sector is downloading data continuously.
17 N. Chen and S. Jordan, "Throughput in Processor-Sharing Queues," *IEEE Transactions on Automatic Control*, vol. 52, no. 2, pp. 299–305, February 2007.

Assume for the moment that the capacity of the system (or link) ρ_c is fixed, then as the system carries more traffic ρ, the user-perceived throughput r decreases.

Rewriting the equation:

$$r = \rho_c - \rho = \rho_c - \rho_c\theta = \rho_c(1 - \theta) = \rho(1/\theta - 1). \quad (11)$$

Whereas traffic $\rho = \rho_c\ \theta$ is linearly dependent on the utilization, the user-perceived throughput $r = \rho_c\ (1 - \theta)$ is linearly related to the complement of the utilization. In other words, UPT is linearly related to the percentage of resources that are unused.

Note that the UPT is a property of the queue used by the radio at a particular sector. The derivation of Equations (10) and (11) utilizes the average time it takes to send the average packet size without regards to which subscriber location the packet is sent. Of great interest is the UPT for each subscriber location in the sector. The UPT should vary from one subscriber location in the sector to another due to changes in average data-link speed. I will state without proof that the UPT for a particular subscriber location is equal to the average data-link speed times the percentage of unused blocks, i.e. $s_i(1 - \theta)$.[18] I have several reasons for not using this definition of UPT. First, if one performs an average of the UPTs using $s_i(1 - \theta)$, you will get $(\Sigma s_i/n)(1 - \theta)$, which does not equal Equation (11). Second, a common operational measurement provided by a mobile network equipment vendor to the MNO is the average UPT for a sector for any hour of the day. Either the equipment vendor will state that sector UPT r is defined by Equation (11), or, as when I analyzed the values for r, ρ and θ from OMs, the value for r is usually within 10% of the value for $\rho(1/\theta - 1)$. The counter-argument for using $s_i(1 - \theta)$ for UPT becomes evident when we define cell-edge UPT. The cell-edge UPT is defined by ordering the UPTs of every subscriber location and noting the UPT that defines the lowest 5th or 10th percentile. I have no problem with this definition, but most mobile network equipment vendors do not provide that statistic. I do have an alternate definition for cell-edge UPT that I will describe later, which can be verified against network OMs.

Another way to look at UPT as defined by Equation (11) is to envision a particular subscriber downloading the same packet size at every subscriber location in the sector. The UPT is the total amount of data the subscriber downloads divided by the total time it took to download the data while visiting each subscriber location in the sector. There is only one UPT number for a sector, which assumes that every subscriber in a sector performs the same routine: i.e. downloads the same amount of data at each subscriber location within that sector.

18 L. Kleinrock, *Queuing Systems, Vol 2, Computer Applications*, Wiley-Interscience, New York, NY, 1976.

Note that the derivation of the UPT formula assumes that there are many users in the sector. If you are the only person in the sector and you decide to use 100% of the resources for downloading some huge file, your UPT will not dive to 0 Mbps. Your UPT will be the speed of the data link. If another user joins you on the link, depending on how much data this user downloads, your download speeds may not decrease (if the second user downloads a small amount) or, in the worst case, you may see your speeds cut in half (if that person is also downloading a huge file). In either case, it is not going to zero, as predicted by the equation. For the equation to work, you need several dozen active users in the sector. In most macro sectors, this requirement is easily met. For scenarios where there is only a handful of active users, you can divide the capacity of the sector ρ_c by the number of active users to get an approximation of the average user-perceived throughput when the link approaches 100% utilization (i.e. $\theta = 1$). This is, in fact, the assumption that the 3GPP consortium makes when it performs simulations in developing new air-interface standards. So, when it performs simulations to demonstrate the capacity of the network using the latest technology, it drives each sector to 100% utilization and shows that UPT doesn't go to 0, because it limits the average number of users to 10 per sector. In a deployed network, if an operator tried to drive network utilization to 100%, UPT would likely go to 0 because the number of subscribers on a congested sector is typically well over 100.

This focus on UPT is a relatively new development, compared to the previous preoccupation with traffic and capacity. For many decades, the telecom industry was focused on increasing the capacity of wireless networks. The subscriber, on the other hand, is more concerned about UPT. Telecom operators wanted the ability to carry more traffic, thus generating more revenue, and the network vendors complied by devising more spectrally efficient technologies. The preoccupation on traffic took focus away from improving the user experience. Increasing spectral efficiency (thus increasing ρ_c) does improve user-perceived throughput, but indirectly. Perhaps it was unfamiliarity or the reluctance to use queueing theory concepts that explains the relative lack of focus on UPT. Perhaps it was the fact that increasing this throughput reduces the amount of traffic that can be carried.

It will be shown later that maximum network capacity usually occur when all base-station radio links are at 100% utilization[19], which means that UPT approaches 0 if there are more than several dozen active users per sector, as was just shown. Although mobile network operators do not want to operate the wireless network at 100% utilization, it

19 There are cases where traffic decreases as radio utilization increases. We will analyze such unusual cases, as they provide insight into the nature of wireless networking.

didn't stop equipment vendors from developing proprietary scheduling algorithms to increase capacity in this congested state. Many vendors tout the superiority of their scheduler design versus other vendors. The above derivation of the UPT equation assumed a simple First Come First Served (FCFS) or First In First Out scheduler design. Is it possible that a more sophisticated scheduler can achieve better performance than FCFS under certain conditions? Perhaps, but it will be shown later that the average link utilization for the typical sector during the busy hour is less than 30%, which means that most radios will only have a single user or no users in the queue. There is nothing for the scheduler to schedule. In the few sectors with significant link utilization, the UPT is rapidly approaching 0 bit/sec. The scheduler may improve the situation by a small amount, but it won't make a congested link appear good. Field measurements also indicate that there is little evidence that advanced scheduler designs have any impact on UPT or traffic[20].

This focus on scheduler design may be more about product marketing and a convenient research topic for academics than for providing a meaningful boost to network performance. Advanced schedulers are required for air-interface standards to meet high spectral-efficiency requirements set by standard development organizations. However, those spectral efficiency numbers are just marketing figures that won't be seen in actual network deployments. Subsequent chapters in this book will derive the capacity of a typical network under various user-perceived throughput constraints, assuming a simple FCFS scheduler design. One could argue that any network capacity estimate arising from this assumption will be on the conservative side.

2.3.4 Wireless Priority Service Analysis

The material in this section is an application of the concepts just discussed, but understanding it is not essential for reading the remainder of this book. It does, however, demonstrate the power of using UPT to determine the cost of providing a service. The analysis doesn't involve any wireless concepts and can be applied to any communication network. Another interesting aspect of the analysis in this section is that it is purely analytic, as calculations do not require the use of a computer algorithm.

Since the earliest days of analog cellular mobile networks, all public mobile air-interface standards support a feature which allows some mobile subscribers to receive higher priority access to the wireless network than other users. These high-priority customers are typically members of emergency services or other government agencies. There is

20 J. Salo and E. Zacarías, "Analysis of LTE Radio Load and User Throughput," *International Journal of Computer Networks & Communications (IJCNC)*, vol. 9, no. 6, pp. 33-45, November 2017.

high demand for communication services during natural and man-made catastrophes. Mobile network operators do not provision the network to handle extraordinary conditions, hence wireless network can become completely congested during these times. Wireless Priority Service (WPS) allows certain groups of users priority access to communication services over the general subscriber base. It is also possible to offer high-priority service to non-government agencies, but mobile network operators seldom do. It is difficult to market a network as the best when some customers are getting better service than others.

This chapter will not provide the absolute cost of transporting a high-priority gigabyte over a communication link. However, if the cost of transporting a non-prioritized gigabyte over a communication link is known, this chapter will analyze how much more it will cost to transport a gigabyte with higher priority.

To simplify analysis, only two classes of service will be assumed. Type 1 will be the higher-priority service associated with WPS, while type 2 will be the service associated with regular non-prioritized traffic.

Packets associated with type-1 and type-2 customers arrive according to an independent Poisson process, with arrival rates λ_1, and λ_2 respectively (measured in packets/second). The average size of the packets for both type-1 and type-2 customers are $1/\mu$, measured in bits. Traffic for both types of service is just the arrival rate λ_i multiplied by the average packet size $1/\mu$ or $\rho_i = \lambda_i/\mu$. The link utilization for both types of service is $\theta_i = \rho_i/\rho_c$. Necessarily,

$$\theta = \theta_1 + \theta_2 < 1, \tag{12}$$

which means that aggregate utilization cannot exceed 1 or 100%. For example, if type-1 traffic occupies 30% of the capacity and type-2 traffic occupies 40% of the capacity, then 70% of the capacity of the channel is used. Similarly, traffic for each type can be summed to produce the total traffic on the link: $\rho_1 + \rho_2 = \rho$.

In this analysis, the pre-emptive resume priority rule is assumed. Pre-emptive resume priority means that when a type-2 customer is in service and a type-1 customer arrives, the type-2 service is interrupted and the server proceeds with the type-1 customer. Once there are no more type-1 customers in the system, the server resumes the service of the type-2 customer at the point where it was interrupted.

Let the **average** time a data packet of type *i* spends in the system as W_i, and the **average** number of type-*i* customers in the system as L_i, where $i = 1, 2$.

For type-1 customers (those associated with WPS), the type-2 customers do not exist. It is as if the type-1 customers have the sector all to themselves. Recalling Equation (9) we have:

$$L_1 = \frac{\theta_1}{1-\theta_1}, \tag{13}$$

$$W_1 = L_1/\lambda_1 = \frac{1/\mu\rho_c}{1-\theta_1}. \tag{14}$$

The total number of customers in the system does not depend on the order in which the customers are served. This number is the same as in the system where all customers are served in order of arrival. Hence,

$$\begin{aligned} L &= L_1 + L_2 = \frac{\theta}{1-\theta} = \frac{\theta_1+\theta_2}{1-\theta_1-\theta_2}, \\ L_2 &= L - L_1 = \frac{\theta_1+\theta_2}{1-\theta_1-\theta_2} - \frac{\theta_1}{1-\theta_1} = \frac{\theta_2}{(1-\theta_1-\theta_2)(1-\theta_1)}. \end{aligned} \tag{15}$$

Applying Little's Law:

$$W_2 = \frac{L_2}{\lambda_2} = \frac{1/\mu\rho_c}{(1-\theta_1-\theta_2)(1-\theta_1)}. \tag{16}$$

The economic cost of supporting WPS will be determined by how WPS affects the UPT of non-WPS customers. Without different priorities, WPS traffic and non-WPS traffic will experience the same KPIs. Hence, it would make sense that both types of traffic be charged the same rate. With priority service, the impact of the addition of 1 GB of type-1 traffic on type-2 traffic KPIs will be greater than the impact of the addition of 1 GB of type-2 traffic, hence the cost of 1 GB of WPS traffic should be greater.

Recall that the average user-perceived throughput r_i is defined as the ratio of the average size of the data packet $1/\mu$ to the average time W_i it takes to transmit that packet, i.e.:

$$r_i = \frac{1/\mu}{W_i}. \tag{17}$$

CASE 1

Analysis without priority queueing. With no type-1 WPS traffic, i.e. only type-2 traffic is present, then:

$$r_2 = \frac{1/\mu}{W_2} = \frac{\lambda_2/\mu}{L_2} = \frac{(1-\theta_2)\,\rho_2}{\theta_2} = (1-\theta_2)\rho_c = \rho_c - \rho_2. \tag{18}$$

Now add type-1 traffic but without priority, i.e. type-1 and type-2 traffic have the same priority. Type-1 and type-2 traffic will have the same UPT:

$$r = r_1 = r_2 = (1-\theta_2-\theta_1)\rho_c = (1-\theta)\rho_c = \rho_c - \rho = \rho_c - \rho_2 - \rho_1. \tag{19}$$

CASE 2

Analysis with priority queueing. As mentioned above, with priority queueing, type-1 traffic behaves as if type-2 traffic was not present at all:

$$r_1' = \frac{1/\mu}{W_1} = \frac{\lambda_1/\mu}{L_1} = \frac{(1-\theta_1)\,\rho_1}{\theta_1} = (1-\theta_1)\rho_c = \rho_c - \rho_1. \tag{20}$$

For type-2 traffic, using:

$$\begin{aligned} r_2' = \frac{1/\mu}{W_2} &= \rho_c(1-\theta_1-\theta_2)(1-\theta_1), \\ &= \rho_c(1-\theta)(1-\theta_1)\,, \\ &= (\rho_c-\rho)\left(1-\frac{\rho_1}{\rho_c}\right), \\ &= \rho_c - \rho - \rho_1 + \frac{\rho\rho_1}{\rho_c}, \\ &= \rho_c - \rho_2 - \rho_1\left(2-\frac{\rho}{\rho_c}\right). \end{aligned} \tag{21}$$

Recall from Equation (19) that, without priority queueing, $r_2 = \rho_c - \rho_2 - \rho_1$. Analyzing Equations (19) and (21), it is easy to notice that ρ_1 is replaced by:

$$\rho_1 \to \rho_1\left(2-\frac{\rho}{\rho_c}\right) = \rho_1 + \rho_1\left(1-\frac{\rho}{\rho_c}\right). \tag{22}$$

Thus, the impact of adding ρ_1 bps of type-1 WPS traffic is equivalent to adding:

$$\Delta\rho = \rho_1\left(1 - \frac{\rho}{\rho_c}\right) = \rho_1(1 - \theta), \tag{23}$$

of traffic. The normalized cost of WPS (i.e. the premium that should be charged for type-1 WPS customers) is:

$$\frac{\Delta\rho}{\rho_1} = \left(1 - \frac{\rho}{\rho_c}\right) = (1 - \theta). \tag{24}$$

Although the result is simple, it is also counterintuitive. Recall that θ is the occupancy rate or the utilization of the channel. According to Equation (24), when the utilization of the channel is high (θ close to 1), the premium is very low (~0%). When the utilization of the channel is low (θ close to 0), the premium is very high (100%). But the whole point of WPS is to give priority to type-1 customers when the channel is close to congestion (high utilization), the scenario in which priority scheduling gives the greatest benefit. However, this scenario, when giving WPS customers higher priority, has the least impact on the UPT of the type-2 customer. When the sector is close to congestion, type-2 customers are seeing very poor throughput. Giving type-1 customers priority isn't going to make the type-2 UPT that much worse. Hence, the premium that should be charged to type-1 customers for degrading the type-2 QoE (Quality of Experience) should be very low.

To make the point clearer, I'll introduce this analogy. Say you live in a large metropolitan area and must take the highway to get to and from the office. When traffic is very heavy, the resulting speed of travel is very slow. As you are sitting in traffic, you hear a siren and see flashing lights coming from behind you. You know that you might have to vacate your lane if you are in the passing lane, or that people in the passing lane may have to merge into your lane to make room for the emergency vehicle. As a low-priority commuter, your commute has become a little worse, but overall you don't adjust your estimate of your arrival time as the average speed of your commute has not changed. The emergency vehicle has little impact on the quality of your commute.

Conversely, when the channel utilization is very low (θ close to 0), the type-1 customers are paying the maximum premium (100%) for having a higher priority. Again, this is not desirable. When the channel utilization is very low, both type-1 and type-2 customers are seeing very good data throughput. It is not likely that a type-1 customer will be willing to pay any premium to get a marginally faster connection, even though this is the scenario where traffic will have the greatest impact on type-2 data throughput (which is still very good even after being slightly degraded by de-prioritization).

To emphasize the point, I will use another analogy. Imagine crossing an international border at an early hour in the morning. From previous experience, you know there is a 2% chance of seeing another traveller being questioned by the border agent when you enter the queue. The other 98% of the time, the queue is empty and you can engage the border agent without waiting. Recently, there has been a policy change, giving first responders priority over other travellers at the border crossing. At this time in the morning, half the people crossing the border are first responders. There is still a 2% chance of the border agent interrogating a traveller when you arrive at the border. While you are being questioned by the border agent, there is a 1% chance of a first responder arriving at the border , thus pre-empting your interrogation and forcing you back into the queue. Before the policy change, there was a 2% chance of someone arriving at the border while you are being questioned by the border agent. After the change in policy, there is still a 2% chance of someone arriving at the border while you are being interrogated, but only half the time the arriving traveller is a first responder, causing you to go into the queue. Hence, instead of waiting in the queue 2% of the time, you are now waiting in the queue 3% of the time, a 50% increase.

With the new policy, the first responder can now bypass the regular traveller when he or she arrives at the international border crossing. However, the person still has to wait 1% of the time since there is a 1% chance that the border agent is interrogating another first responder. This is still a 50% improvement from the time when he or she couldn't pre-empt regular travellers. Some would say that the improvement is only 1%. Instead of not having to wait in a queue 98% of the time, with priority access the first responder does not have to wait 99% of the time, a 1% improvement. It is questionable whether the first responder is willing to pay the premium (twice the price) for such a small improvement in wait times.

To gain further insight, a new factor will be analyzed. This will quantify the utility of the WPS prioritization feature for type-1 customers. The utility of the WPS feature will be based on the amount of improvement in user throughput when the feature is turned on. Using Equations (19) and (20), the improvement of adding priority to type-1 user-perceived throughput is:

$$\begin{aligned} \Delta r_1 &= r_1' - r\,, \\ &= \rho_c - \rho_1 - (\rho_c - \rho_2 - \rho_1) \;=\; \rho_2. \end{aligned} \tag{25}$$

The increase in throughput is equal to the amount of type-2 traffic that is being disregarded due to the prioritization of type-1 traffic (recall that for type-1 customers, type-2 traffic isn't even present). Referring back to the highway commute example and looking at the situation from the first-responder point of view, the biggest impact of having priority access is when the highway is congested. In an ideal world, having priority access is like having the highway devoid of commuters.

The most likely scenario in which WPS is to be deployed is when type-1 traffic represents the minority of the total traffic. It doesn't make sense to give everyone priority access since we are back to the situation where everyone has the same priority. I will analyze two cases. In Case 1, type-1 users represent a very small percentage of the total traffic. In Case 2, type-1 users represent a larger portion of the total traffic.

CASE 1 (Very Low Type-1 Traffic)

To assess the effectiveness of WPS when type-1 traffic is low, let's analyze the relative deterioration of type-1 UPT when priority access is not employed. The maximum possible gain in UPT occurs when the channel is completely congested with type-2 traffic (i.e. $\rho_2 = \rho_c$ and ρ_1 is minimal). In this situation, without priority access, the user-perceived throughput is 0 for both type-1 and type-2 subscribers (i.e. $r_1 = r_2 = 0$). When priority access is invoked, the UPT for type-1 users goes from 0 to $(\rho_c - \rho_1) \approx \rho_c$ if ρ_1 is minimal. Using the results of Equation (25), we define the loss in type-1 UPT in the absence of WPS as:

$$\frac{\Delta r_1}{r_1'} = \frac{\rho_2}{\rho_c} = \theta_2 \,. \tag{26}$$

Hence, in the absence of WPS, type-1 UPT loss is equal to type-2 traffic utilization. When type-2 traffic is very high (θ_2 close to 1), the benefit (or utility) of WPS prioritization is very high, since without it, type-1 UPT loss is close to 100%. When type-2 traffic is very low (θ_2 close to 0), the benefit (or utility) of WPS prioritization is very low since the deterioration in type-1 user throughput without WPS is minimal.

CASE 2 (General Type-1 Traffic Levels)

We are going to define two ratios that will provide some additional insights. The first ratio gives the gain in type-1 user-perceived throughput, i.e. $\Delta r_1/r_1$, using Equation (19) and (25):

$$\frac{\Delta r_1}{r_1} = \frac{\rho_2}{\rho_c - \rho} = \frac{\theta_2 \rho_c}{\rho_c - \theta \rho_c} = \frac{\theta_2}{1 - \theta} \,. \tag{27}$$

Which means that to get the highest percentage increase in UPT, you need to have both high type-2 traffic utilization as well as high total traffic utilization.

The second ratio gives the deterioration in type-1 user-perceived throughput if WPS is not employed. This is a generalization of the analysis outlined in CASE 1. Using Equation (25):

$$\frac{\Delta r_1}{r_1'} = \frac{\rho_2}{\rho_c - \rho_1} = \frac{\theta_2}{1 - \theta_1}. \quad (28)$$

Equations (27) and (28) are related since $(1 + \Delta r_1 / r_1)(1 - \Delta r_1 / r_1') = 1$, a relationship that can be verified via simple substitution.

In all of the above cases, the effectiveness of WPS increases with higher type-2 traffic utilization. As we will later see, the average sector utilization is approximately 30% in a busy cluster of cell sites. At this level, the benefits of WPS are low and the relative costs are high. The solution is to only activate this feature when sector utilization is high (e.g. greater than 70% sector utilization). As this occurs relatively rarely, and the relative cost of providing WPS is low at these levels, it appears that WPS should not be an expensive feature to implement.

2.4 Frequency Reuse

With CDMA (Code Division Multiple Access), HSPA (High-Speed Packet Access), LTE (Long-Term Evolution) and 5G NR (New Radio), the concept of universal frequency reuse or N=1 frequency reuse is fairly well established and exclusively adopted. These air-interface standards are designed to handle poor signal-to-noise ratio conditions arising from severe co-channel interference when all radios share the same spectrum. Other air-interface technologies, such as Wi-Fi and microwave backhaul, use much higher-frequency reuse plans, which means that radios only use a portion of the spectrum band. Both technologies require much higher signal-to-noise ratios to operate. In the case of microwave backhaul, careful frequency planning must be used since very little error detection/correction is employed due to strict latency constraints.[21] I mention this because someone may claim that a certain technology using a higher-frequency reuse plan supports a much higher spectral efficiency. For example, some microwave backhaul links can support spectral efficiencies of 10 to 20 bit/sec/Hz compared to LTE, which in actual deployed networks supports less than 1 bit/sec/Hz. To support such a high spectral

21 There is a trade-off between latency and capacity. Microwave backhaul achieve microsecond latencies while LTE and other mobile wireless interface technologies have latencies measured in milliseconds. LTE sacrifices latency by using time-consuming channel-coding algorithms (i.e. Turbo codes) to achieve higher capacity.

efficiency, a very high signal-to-noise ratio is required. For example, if the microwave spectrum band is 200 MHz, only a small fraction of that band can be assigned to a given link (10 MHz for example). To facilitate comparison between different design choices, perhaps it is better to quote spectral efficiency in terms of capacity of the radio divided by the total bandwidth of the spectrum band. In the above example, the spectral efficiency is only 0.5 to 1 bit/sec/Hz, since the microwave radio uses only $1/20^{th}$ of the spectrum band.

The model I developed can analyze the impact of frequency reuse on spectral efficiency and UPT. Although CDMA inherently uses N=1, with OFDMA (Orthogonal Frequency-Division Multiple Access) there was some uncertainty whether to use N=3 or N=1. OFDMA-based WiMAX was mainly deployed using N=3, while LTE used N=1. The benefit of N=3 is that users experience much higher SINR (signal-to-interference-plus-noise ratio), about 10 dB higher than N=1. However, the higher SINR did not compensate for the fact that only one-third of the spectrum was available for each active user. Given the same amount of spectrum, the tool I developed showed that N=1 supported approximately 50% more traffic for a given user-perceived throughput versus N=3. Since N=1 is superior to other higher frequency reuse plans, all further analysis will assume that universal frequency reuse is being used.

2.5 Radio Frequency Propagation Model

Stating the obvious, the further the user is from the cell site, the weaker the signal received by the user. Path loss is a measure of the decrease in signal strength due to distance. It would be nice to be able to predict the path loss exactly for any situation, but this is highly difficult due to many factors, such as terrain, obstructions, reflections, etc. The best that can be done is to use a model to predict the median path loss based on factors such as distance, frequency, and antenna heights.

Figure 6: *Physical model of free-space propagation.*

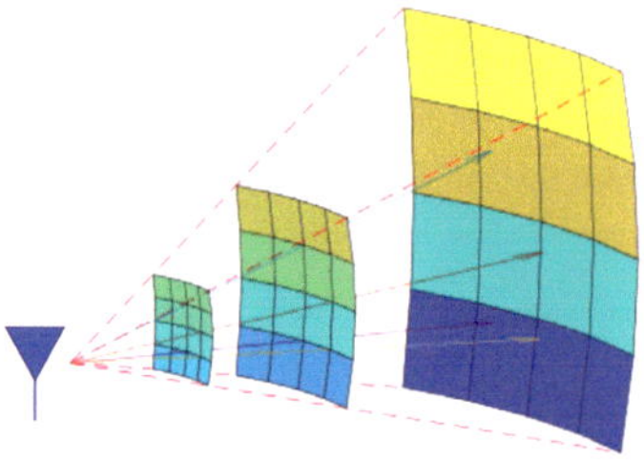

At a minimum, electromagnetic signal strength will diminish as it propagates from the transmit antenna to the receive antenna even if there are no obstacles to impede RF propagation. The amount of attenuation is called the free-space path loss, and it increases by the square of the distance covered ($distance^2$ or d^2). The free-space path loss assumes that both the transmit and receive antennas are isotropic radiators, which means that the antennas transmit and receive energy equally in all directions.

If an antenna transmits RF power in all direction equally, then one can visualize a sphere—with the transmit antenna at its center—with the amount of RF power at each point on the surface being equal. (See Figure 6.)

The total amount of power crossing the entire surface doesn't depend on the size of the sphere, but the amount of power crossing a patch of a specific size—say a square 1 meter by 1 meter—does. The area of the sphere is simply $4\pi d^2$, where d is the radius. The RF power density traversing the sphere is the amount of RF power traversing that 1-meter square and is simply the transmit power divided by the area of the sphere, or $P/4\pi d^2$. Interestingly, the amount of power traversing the sphere does not depend on the frequency or wavelength of the radiator. To derive the free-space path loss, the capture area of the isotropic receive antenna must be specified, which is a function of the wavelength of the RF signal. The derivation of the effective size or capture area of the isotropic radiator is complicated, but the resulting formula is very simple: $A_{eff} = \lambda^2/4\pi$, where λ is the wavelength of the RF signal. The higher the frequency, the smaller the wavelength (since $\lambda = c/f$ where c is the speed of light and f is the frequency), and the smaller the size of A_{eff}. The power of the received signal is simply the effective size of the isotropic receive antenna (A_{eff}) multiplied by the RF power density or $P\text{x}A_{eff}/4\pi d^2 = P(\lambda/4\pi d)^2$. The formula indicates that the power of the transmit antenna should be multiplied by the gain factor $(\lambda/4\pi d)^2$ to get the power at the receiving isotropic antenna. Free-space path loss is defined as the reciprocal of the gain factor or $(4\pi d/\lambda)^2$. The path-loss exponent defines how fast the path loss increases with distance and, in the case of free-space path loss where path loss increases as the square of the distance, the path-loss exponent is 2. As a matter of convenience, the free space path loss is often expressed in dBs.

$$FSL(dB) = 10\log\left(\left(\frac{4\pi d}{\lambda}\right)^2\right). \quad [dB] \tag{29}$$

$$FSL(dB) = 20\log\left(\frac{d}{1000}\right) + 20\log f_c + 32.45. \quad [dB] \tag{30}$$

Where d is the distance in meters and f_c is the frequency in MHz.

When the distance between the transmit and receive antenna increases by a factor of 10, the free-space path loss increases by 20dB or by a factor of 100. Free space propagation does not occur very often in mobile communication. Even if there is line-of-sight between the transmitter and receiver, the transmit signal will reflect off the ground and will partially cancel the direct signal at the receive antenna. The two-ray ground reflection model is a much more relevant path-loss model in mobile communications than the free-space path-loss model, and simple calculations will show that the path-loss exponent is 4. This means when the distance between the transmit and receive antenna increases by a factor of 10, path loss will increase by 40 dB or by a factor of 10,000. The path-loss equation in dB for the two-ray ground reflection model will have the form 40log(d) + constant.

The free-space path loss model and two-ray ground reflection model are based on a theoretical explanation of how RF propagates in the physical environment. Another method of developing a RF propagation model is to empirically collect data from an actual wireless network. Most empirically based models are variants of ones originally developed by Okumura and Hata. 3GPP has adopted the COST Hata model:

$$PL = (44.9 - 6.55\log h_b)\log\left(\frac{d}{1000}\right) + 45.5 + (35.46 - 1.1h_m)f_c - 13.82\log h_b + 0.7h_m + C + F. \quad [dB] \tag{31}$$

Where:

- h_b is the height of the base-station antenna in meters
- d is the distance between the base station and mobile in meters
- h_m is the height of the mobile antenna in meters
- f_c is the frequency of the carrier in MHz
- C is a correction factor
  - C = 0dB for suburban and rural
  - C = 3dB for urban environments
- F is the term associated with log-normal fading.

The path-loss exponent is γ = (44.9 – 6.55 logh_b)/10. Although the COST Hata model has many terms, it is quite easy to use. When the log-normal term F is set to 0, the formula will give the median path loss in dBs.

Another possibility is to modify the COST Hata model using field measurements. Regression is used to find the unknown constants in the equation below.

$$PL = 10\gamma\log\left(\frac{d}{1000}\right) + C_0 + C_1\log f_c - C_2\log h_b + C_3h_m + F. \ [dB] \tag{32}$$

One possibility is to use values from Equation (31) for C_1, C_2 and C_3 and only calculate C_0 and γ. Note that γ in Equation (32) is not a function of the base-station antenna height as suggested by (31). There are two reasons for this simplification: A) when performing field measurements, it is unlikely there will be a sufficient number of measurement campaigns with base stations of different heights to make γ a reliable function of base station height; and B) analysis is simplified when γ is independent of the base-station antenna height.

Radio frequency propagation models that are based on the results of empirical measurements generally have the following equation for the median path loss:

$$Median\ Path\ Loss = 10\gamma \log(d) + C_0. \quad [dB] \qquad (33)$$

Where d is distance in meters, γ is the path-loss exponent and C_0 is some constant. The median path-loss Equation (33) is another part of the foundation of the method I've developed. As simple as this equation appears, quite often it can be made simpler if C_0 is much smaller than $10\gamma\log(d)$ and can be ignored.

2.5.1 Fading

For the sake of completeness I should spend a little time explaining some of the complexities associated with the fading term F in Equations (31) and (32). Both equations without the F term will give median path loss—a specific number—as a function of the values given to the other variables in the equation. However, it is understood that path loss is a random variable and the purpose of F is to allow the path-loss equation a range of values around the median path loss —not just one number. The reason we can't assign a single number for path loss is because of a phenomenon called fading that makes calculating the exact path loss for a particular RF link extremely difficult. It is like, for example, predicting the exact salary of an individual college graduate 10 years in the future, even though there are statistics that give the median salary of college graduates 10 years after receiving their diplomas.

There are two components that comprise the path-loss equation, but fortunately for us only one matters. There is long-term or log-normal fading, which causes average signal-strength variations over long distances, and short-term or Rayleigh fading which causes signal-strength variations on the order of centimeters. Long-term fading arises from the effects of shadowing caused by large objects blocking the direct RF path. Short-term fading is a result of the same RF signal arriving at the antenna from many directions due to a local cluster of scatterers reflecting the RF signal in the direction of the receiver. The figure below is illustrative of the process that give rise to long-term and short-term fading.

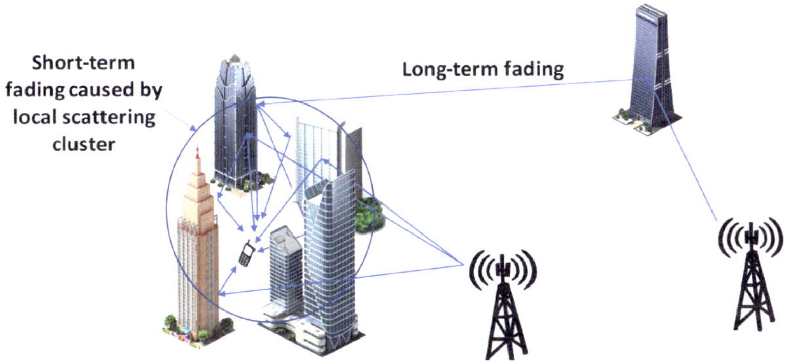

Figure 7: *Causes of long-term fading (Log-normal shadowing) and short-term fading.*

Both types of fading cause variations in the amplitude of the signal. The figure below shows the variations of signal amplitude caused by short-term fading.

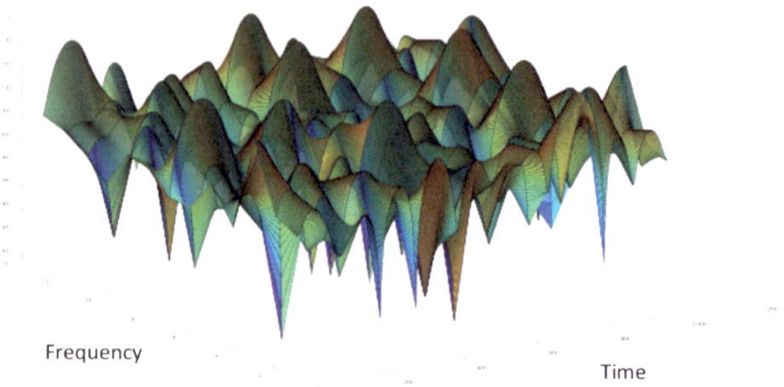

Figure 8: *Short-term or Rayleigh fading.*

For wideband signals (which are the only signals of interest in the age of broadband data communications), the average signal strength over the local cluster area (see Figure 7) is fairly constant, even though components of the signal are attenuated in frequency, space and time due to short term-fading. It is the responsibility of those designing the physical layer to mitigate and even exploit the effects of short-term fading in the design of the RF air-interface. MNOs should focus on long-term fading as it accounts for differences in the average signal strengths amongst local cluster areas. MNOs can avoid dealing with the impacts of short-term fading by using Figure 4 to determine data-link speeds for a given SINR, as the effects of short-term fading are included in the graphs.

Long-term fading is usually modeled by attributing the random variable '*F*' to the log-normal probability distribution with a standard deviation ranging from 6 to 12 dB.

2.6 Thermal Noise

The Shannon-Hartley equation indicates that the data rate of a channel is controlled by the signal-to-noise ratio. In cellular communication, the noise sources are other radios transmitting on the same frequency band for which the receiver in question is not the intended target. This is referred to as co-channel interference. But even if co-channel interference is non-existent, there is a limit to how weak a signal can become and still be received due to thermally generated noise in the environment. The thermal noise power is given by kTB, where k is Boltzmann's constant (1.38×10^{-23}), T is the absolute temperature of the load (typically 290° K), and B is the measurement bandwidth. If B is set to 1 Hz, then 10log(kTB) = -204 dBW or -174 dBm. The original North American analog cellular technology (AMPS) had a channel bandwidth of B = 30 kHz. The thermal noise floor for an AMPS receiver is -129 dBm.

The front-end of the receiver also injects additional noise, which degrades the signal-to-noise ratio. The noise figure of the receiver front-end is an indication of how much it raises the noise floor. The noise figure of a typical hand-held device is approximately 7dB. This means that instead of the noise floor being -129 dBm for an AMPS 30 kHz receiver, it is -122 dBm. A typical FM receiver requires a signal-to-noise ratio of around 4 dB to avoid static. This means that the weakest signal that an AMPS handset can demodulate without static is -122 + 4 = -118 dBm.

2.7 Minimum-Required Signal Strength

The power of the RF signal at the input of the receiver can be calculated using the following formula:

$$\begin{aligned} &Received\ Signal\ Strength\ (dBm) \\ &\quad = Transmit\ Power\ (dBm) + Gains\ (dB) \\ &\quad - Path\ Loss\ (dB) - Misc\ Loss\ (dB) \end{aligned} \tag{34}$$

The Path-Loss (dB) term in Equation (34) is specified by the radio-frequency-propagation model (Equation (30) to Equation (33)) and the Misc Loss term includes losses due to building penetration, cable, connector, body and others. The main sources of gain are antenna gains at the transmitter and receiver. Transmit antennas achieve their gains by concentrating power in a certain direction (typically in the direction of the receiver) while receive antennas achieve their gains by increasing sensitivity in a particular direction (typically in the direction of the transmitter).

Although in theory it is possible to receive signal levels lower than the power of noise[22], for the sake of increasing capacity and limitations imposed by LTE it is recommended that the received signal strength be at least as high as the power of noise. As specified by the Shannon-Hartley equation, the higher the received signal-to-noise ratio (SNR), the faster the data-link speed. Coverage is defined as the area where the received signal strength is higher than a certain value, which we will call the minimum-required signal strength (minSS). Since it is difficult to predict the amount of co-channel interference that may be present, the minimum-required signal strength is specified with respect to the thermal noise floor. Recall that thermal noise, given by kTB, is primarily a function of B (i.e. spectrum bandwidth), since k is the Boltzmann's constant and T is nominally set to 290° K. Also recall that the receiver front-end injects noise, thus raising the

22 From *WCDMA for UMTS: HSPA Evolution and LTE* (H. Holma & A. Toskala, John Wiley & Sons, 2010), LTE can demodulate signals with a SNR of -7 dB.

thermal noise floor. Once the thermal noise floor has been established, the specified minimum-required SNR (SNR_{min}) will determine the minimum-required signal strength. It will be explained later that the minimum-required SNR is specified in many situations since we may not know what value it should be set to. In any case, once the thermal noise floor and the minimum-required SNR have been established, the minimum-required signal strength will also be known.

2.8 Wireline Analogy

Many of the concepts I presented in this chapter can be applied to wireline networks. Wireline networks are much easier to model because crosstalk (the wireline version of co-channel interference) is much lower than the co-channel interference found in wireless networks. In many cases, crosstalk can be ignored, and the network analysis will be in its most simplified form (See Figure 9).

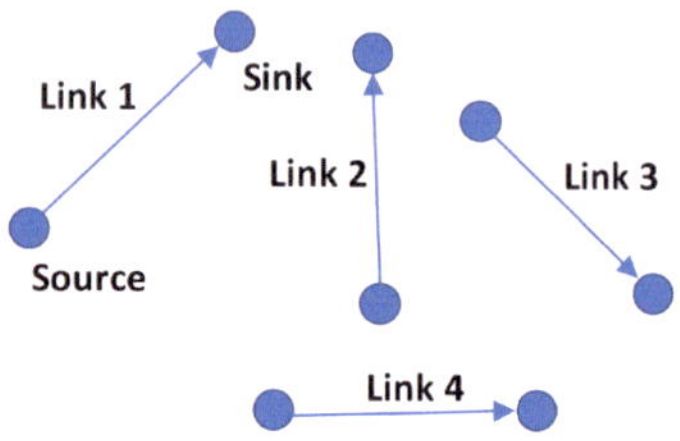

Figure 9: *Hypothetical four-link wireline network.*

To simplify the analysis even more, we will assume that the data links are unidirectional, and the link speeds are equal and constant. To make use of the concepts in this chapter, the source node in the above figure is equivalent to a base station, and the sink node corresponds to a collection of fixed subscribers. It is assumed that there are several dozen subscribers at each sink node sharing the fixed link in time-division duplexing manner. To handle the inevitable contention, a queue employing a First Come First Served scheduler is used.

Calculating the maximum capacity of the network is straightforward.

Figure 10: *User-perceived throughput vs traffic for each of the four links.*

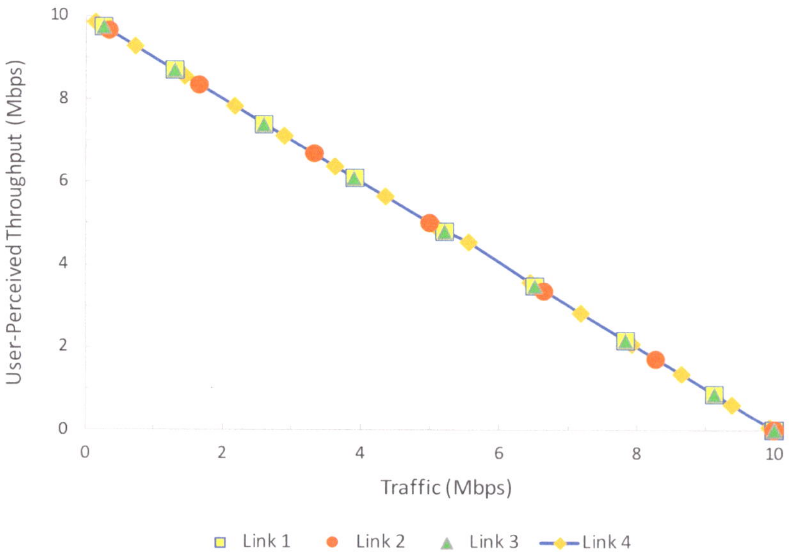

Simply sum the link speeds. If each link supports ρ_c = 10 Mbps, then the above network can support 40 Mbps. At maximum capacity (utilization $\theta = 1$), the UPT as calculated by Equations (10) or (11) is 0 Mbps. Understandably, the wireline operator would never operate its link at maximum capacity. The obvious solution is to operate each link at a lower utilization, such as 70% to 80%. Figure 10 shows the trade-off between traffic and UPT for the four links.

The relationship between user-perceived throughput (r) and traffic (ρ) for each of the four links is identical and can be expressed by the linear equation $r = \rho_c - \rho$ where ρ_c is capacity of the link (10 Mbps). For this example, let's assume the wireline operator chooses θ = 80% as the maximum utilization for any particular link. At $\theta = 0.8$, the link traffic is $\rho = \rho_c\theta = 10 * 0.8 = 8$ Mbps and the user-perceived throughput is $r = \rho_c - \rho = 10 - 8 = 2$ Mbps. Hence, one might conclude that the capacity of the network is now 4 * 8 = 32 Mbps.

However, there is one more twist. Despite its best efforts, the wireline operator couldn't equalise the number of subscribers on each link. Assume that each subscriber downloads the exact same amount, and the number of subscribers at each sink node is given in the table below.

Link Number	Number of Subscribers	Traffic Percentage
1	180	26.1%
2	230	33.3%
3	180	26.1%
4	100	14.5%

Table 4: *Number of subscribers at each sink node.*

We need to recalculate the capacity of the network assuming the above subscriber distribution. As subscriber traffic rises, it appears that the first link to reach capacity is Link 2. Using the same maximum link utilization constraint of θ = 80%, the amount of traffic Link 2 can support is $\rho_2 = \rho_c\theta = 10 * 0.8 = 8$ Mbps. Since Link 2 has one-third of the subscribers (and hence traffic), the total network traffic when Link 2 is at capacity is 8 / 0.333 = 24 Mbps. Notice that network capacity is only 60% of the sum of the link speeds.

The four-link wireline network is so simple that it can be analyzed without the aid of computer algorithms. Unfortunately, with wireless networks, traffic on one link will create co-channel interference with other links, lowering the data-link speeds. With only four links, it might be possible to put each wireless link on its own frequency band and not have to deal with co-channel interference.[23] If that is the case, the analysis in this section can be used to determine the performance of

23 We are going to examine another interesting scenario where interference does not affect network KPIs, not due to its absence but because of its invariability.

the network with some slight modifications. Although there is no co-channel interference, you still must deal with the limitations imposed by thermal noise to determine the data-link speeds and the capacity of the channel. These are much easier to deal with and may not require the use of computer algorithms. All mobile network operators have finite spectrum resources and, as traffic requirements grow, new wireless links must reuse occupied spectrum. This analysis, which is the subject of the next chapter, is much more complicated and requires the use of computer algorithms.

The graph below shows the trade-off between UPT and network traffic.

Figure 11: *User-perceived throughput vs traffic for wireline network.*

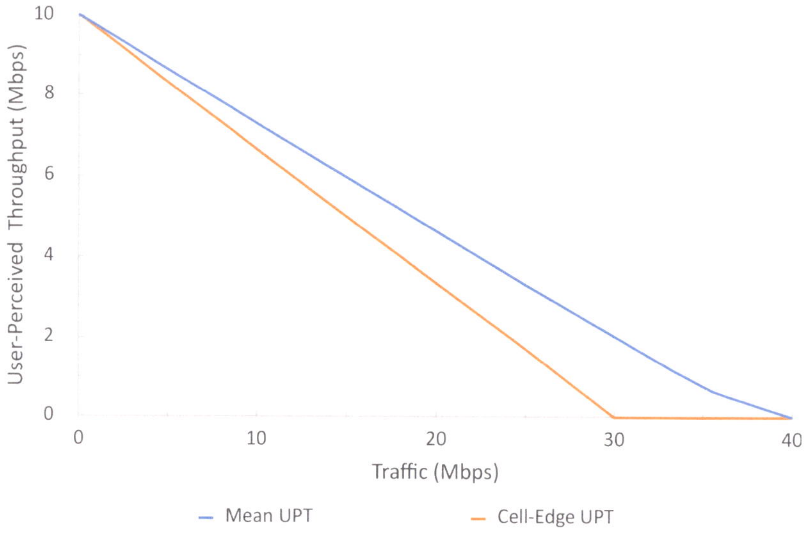

The cell-edge UPT is in the fifth percentile for all subscribers. In other words, the UPTs of all subscribers are ordered and the UPT representing the fifth percentile slowest speed is noted. In this case, all the users in Link 2 have the same user-perceived throughput and reside in the lowest 33rd percentile. Hence, the cell-edge UPT is just the UPT of Link 2. Referring to the above figure, when the UPT of Link 2 reaches 2 Mbps, the network traffic is 24 Mbps, which is three times the traffic on Link 2 (i.e. traffic on Link 2 is 8 Mbps).

Another interesting graph shows link traffic versus network utilization. (See Figure 12.)

Below 75% network utilization, the relative amount of traffic on each link is equal to the relative number of subscribers on each link. For example, when Link 4 is transmitting 100 kbps (100 subscribers each transmitting 1 kbps), Link 1 and 3 are each transmitting 180 kbps and Link 4 is transmitting 230 kbps. This is the area of operation of greatest interest, as no links are in congestion. At 75% network utilization, link 2 reaches 100% link utilization as traffic cannot exceed 10 Mbps. The ratio of traffic for links not in congestion is equal to the ratio of customers supported by each link as each subscriber downloads the same amount of data. For example, if Link 4 is transmitting 5 Mbps (100 subscribers each transmitting 50 kbps), Link 1 is transmitting 9 Mbps (180 subscribers each transmitting 50 kbps). At 100% network utilization, all links are fully utilized and carry the same amount of traffic (10 Mbps). The significance of this chart will become more apparent when we examine more complicated wireless scenarios where data-link speeds are affected by co-channel interference from neighboring cell sites.

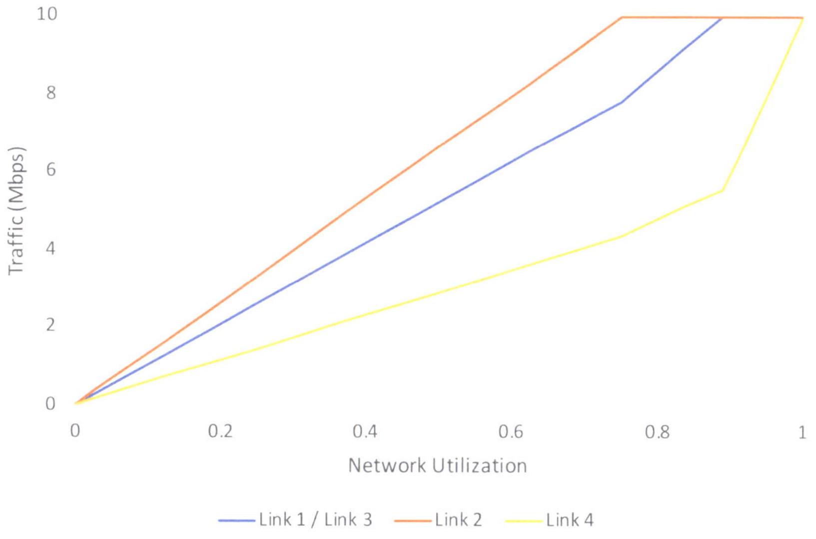

Figure 12: *Link traffic versus network utilization.*[24]

24 Note that graph for Link 1 and Link 3 are identical since they have the same number of subscribers.

3 | The Stochastic Approach

From an MNO point of view, one big problem with the results of most simulations performed by academia and industry is the assumption that traffic is uniformly distributed geographically, causing every sector in the network to carry the same amount of traffic. However, in actual networks, some sectors carry more traffic than others, significantly more in some cases. (See Figure 1.) Also, these same simulations attempt to maximize traffic by stipulating that every radio is continuously transmitting data, a state of network congestion rarely seen due to engineering programs designed to prevent this from occurring. Lastly, fancy scheduling algorithms designed to exploit full buffer congestion are used to increase spectral efficiency, while acceptable UPT is maintained by artificially limiting the number of users in the simulation to less than a dozen per sector. Due to the inappropriateness of these assumptions, MNOs can use the results of simulations as a guide, but they must ultimately devise their own method to dimension networks or evaluate new wireless strategies.

Engineers prefer to use formulas whenever possible, as evidenced by the number of equations presented thus far. In the days when voice was the only service offered by MNOs, engineers used the Erlang-B formula to calculate the number of circuits required to support the anticipated amount of voice traffic. Unfortunately, with the advent of CDMA and universal frequency reuse, the application of formulas is generally more difficult, except for one particular scenario. We know from the previous chapter that the capacity of the sector is given by Equation (6). Under the condition of universal transmit buffer congestion, when every radio in the network is transmitting 100% of the time, the SINR at every point is constant. Data-link speed s_i at every point is also constant, thus, calculating sector capacity ρ_c from Equation (6) becomes straightforward. Although it is recommended that a computer program be used for the computation, manual calculation is possible. MNOs need not resort to Monte Carlo simulations. UPT is also easy to calculate, as it is 0 Mbps during transmit buffer congestion.

However, computation becomes much more difficult under the more prevalent and desirable scenario of negligible radio buffer congestion. The amount of interference at each point will be constantly varying, causing the SINRs and data-link speeds to change continuously. The amount of data each sector downloads is a function of how much data the other sectors are downloading, since any data transmission will create interference in other sectors. Assume there is a need to analyze a network that has a traffic distribution resembling the histogram in Figure 1. The amount of sector traffic can be defined by using $\rho = \rho_c\theta$ (Equation (7)), which mean that ρ_c and θ are required for each sector. Is there a deterministic solution to that problem? Perhaps convex optimization techniques could be brought to bear, but the solution—if one exists—won't be discussed here. I'm more interested in using a stochastic algorithm that exploits randomness to obtain a solution for Equation (6) under non-full-buffer scenarios.

Aside from the analytical challenges, co-channel interference has a bad reputation, as it often gets blamed for poor network performance. But in the wireless world, where universal frequency reuse is required to maximize spectral efficiency, networks should be co-channel interference limited to maximize capacity and avoid coverage holes. Co-channel interference will always be present with universal frequency reuse; adding thermal noise will not improve the situation.

Developing a computer algorithm to solve the problem of analyzing co-channel interference is only half the battle. Even if you develop a Monte Carlo simulator, there is no guarantee that the effort will produce meaningful insights in a timely manner. Before a simulation can be executed (which can take days to weeks depending on the algorithm), each of the following parameters must be configured:

- radio transmitter power
- receiver noise figure
- cell-site location and spacing
- frequency band of operation
- spectrum bandwidth
- antenna patterns
- cell tower height
- distribution of subscriber locations
- network resource utilization
- radio frequency path-loss model assumptions
- scheduling algorithms
- advanced antenna technology assumptions.

With at least a dozen parameters that can be adjusted, a team of data scientists could be running countless simulations for a very long time, generating massive amounts of data and making analysis difficult, if not impossible. Well, that would be the case, if not for the method to be developed in this chapter, which avoids the problem of repeatedly simulating the same scenario even though the parameters appear, on the surface, to be different.

3.1 Cell Site and Mobile Location

MNOs have the option of analyzing a wireless network with actual base-station locations and cell-site configurations or using a generic cell-site plan. (See Figure 13.) Perhaps using actual cell-site location and configuration information is necessary from an operational point of view; if you are the engineer in charge of optimizing a particular cluster of cell sites in a certain part of town, you should include all this information before changing or augmenting the network. But analyzing an actual wireless network makes little sense unless you also have a general idea of where traffic is located. You could use various OMs such as timing advance, Resource Block utilization, total traffic, etc., along with cell-site location and configuration information to estimate where the users are located using a computational technique called regression. Using regression for the purpose of locating traffic is not trivial and books have been written on this subject. For generating insights for the purposes of devising strategic network initiatives, using actual site locations is not mandatory. There is a risk when using actual base-station locations and configuration information that the results might be specific to that particular network. If actual base-station placement and configuration are not used, they can be treated as parameters that can be assigned predetermined values which can be adjusted to observe their effects on network KPIs.

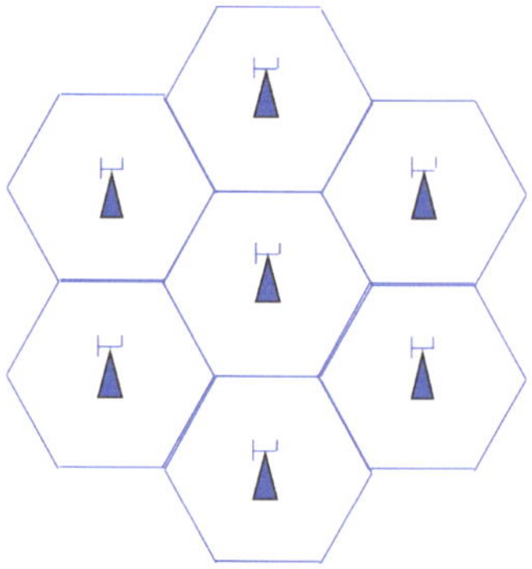

Figure 13: *Hexagonal grid pattern.*

I've already mentioned some of the shortcomings of using heuristics (Chapter 1) and the unrealistic assumptions adopted by industry researchers when performing Monte Carlo simulations. When looking for an alternative method for dimensioning networks or devising wireless strategies, one cannot simply ignore the copious amounts of available OM data which is considered a treasure trove in the minds of some network planning engineers and a prerequisite for those intent on using the heuristic method. The weakness of the heuristic method is it ignores the plethora of information in Chapter 2. In my approach I did not use OMs to guide wireless strategies directly, but to help develop an abstract model based on tunable parameters. Instead of examining OMs, the method I am proposing involves studying how modifying the model's parameters affect the simulated OMs. The model can be configured to generate OMs that have the same statistical properties of OMs found in actual networks. After studying the effects of changing various parameters, one can ignore the OMs from actual networks and rely on the model instead. The benefits of using this abstract approach instead of examining actual network OMs will become more evident when we analyze various network configurations.

To ease into the development of the method I've designed for wireless networks analysis, I will start with a very simple example. It will consist of four omni-directional cell sites equally spaced in a square grid (nominally small cells, although the original AMPS cell sites in 1983 were also omni-directional). Analyzing such a small network is not typical in industry or academia, but it can get unwieldy to analyze results when the network consists of dozens of base stations. The main goal of this chapter is to describe a method that will calculate network capacity and other KPIs, a task more easily accomplished by limiting the size of the network.

As with Chapter 1, the focus will be on the downlink portion of the radio link, since this is where the majority of wireless traffic occurs. This is also the link that people care more about when performing data speed tests on their smartphones. Rest assured, performing wireless network analysis for the uplink is not more complicated.

The first thing I establish is a grid where the mobile subscribers can be located. Although I will specify numbers that can be interpreted as distances, don't be too concerned if some of those seem a little off. They are arbitrary and the scale can be adjusted easily so that any inter-cell site distance can be analyzed without changing the actual cell-site locations on the grid. The following MATLAB statement establishes a grid with a spacing of 25 units (each unit is nominally 1 meter but can represent any other spacing, depending on the scenario).

```
[xcoord, ycoord] = meshgrid(0:25:3500,0:25:3500);
```

The meshgrid command is used to generate two arrays containing the x- and y-coordinates at each position in the rectilinear grid. The above MATLAB command returns two 141 × 141 matrices. The `xcoord` matrix defines the x-coordinates and the `ycoord` matrix the y-coordinates at each position in a 141 × 141 grid. Below is the first 5 x 5 segment of the `xcoord` matrix.

0	25	50	75	100
0	25	50	75	100
0	25	50	75	100
0	25	50	75	100
0	25	50	75	100

Table 5: *The first few entries of the xcoord matrix. xcoord is a 141 x 141 matrix.*

The following is the first 5 x 5 segment of the `ycoord` matrix.

0	0	0	0	0
25	25	25	25	25
50	50	50	50	50
75	75	75	75	75
100	100	100	100	100

Table 6: *The first few entries of the ycoord matrix. ycoord is a 141 x 141 matrix.*

Together, these two matrices form the grid where the mobile subscribers are to be potentially placed. The figure below shows the lower 5 x 5 portion of the grid.

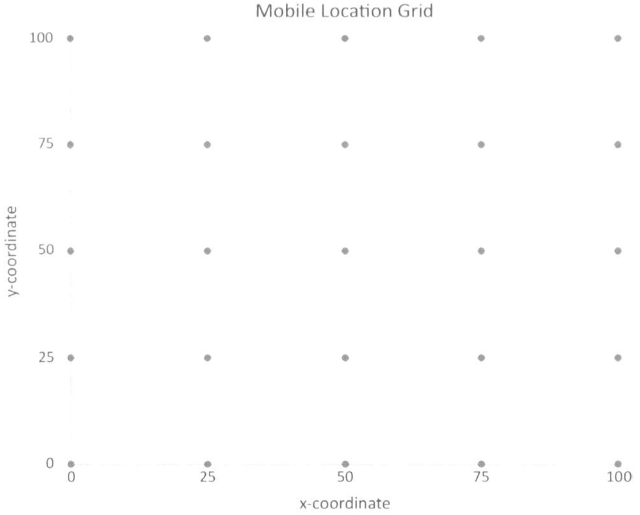

Figure 14: *Grid where mobile subscribers are to be placed.*

Some may argue it is more realistic to place the mobile subscriber anywhere in the 3,500 x 3,500 area. When I run out of points in the grid to place mobile subscribers, I will place them off the grid. If there are sufficient grid points to support the desired number of subscriber locations, I found there was nothing to be gained by placing the mobile subscriber anywhere in the continuum. The main purpose of placing mobile subscribers is to fix the mobile distances to all cell sites in the hypothetical network and then calculate the RF path loss. The median path loss is a function of distance, but a large random component is added (caused by log-normal fading) that masks any changes in RF path loss from small variations in distance. There is an advantage in using a grid for mobile locations when non-uniform mobile location placements are considered. However, in this first example, mobile users will be placed uniformly and randomly so that there are no clusters or concentration of mobile users in one particular area. (See Figure 17.)

Now that the possible mobile locations are determined, it is time to develop a grid for the possible cell-site locations. I could employ the same grid used to place the mobile subscribers, but then I would have to account for the possibility that a mobile subscriber and cell site could occupy the exact same location. The RF path loss equation cannot accept a distance of 0 meters. There are several solutions to this problem, but I've decided to define a new grid that is offset from the possible mobile user locations. The following MATLAB command establishes this grid:

```
xb = xcoord + 25/2;
yb = ycoord + 25/2;
```

The figure below shows the possible location of the cell sites.

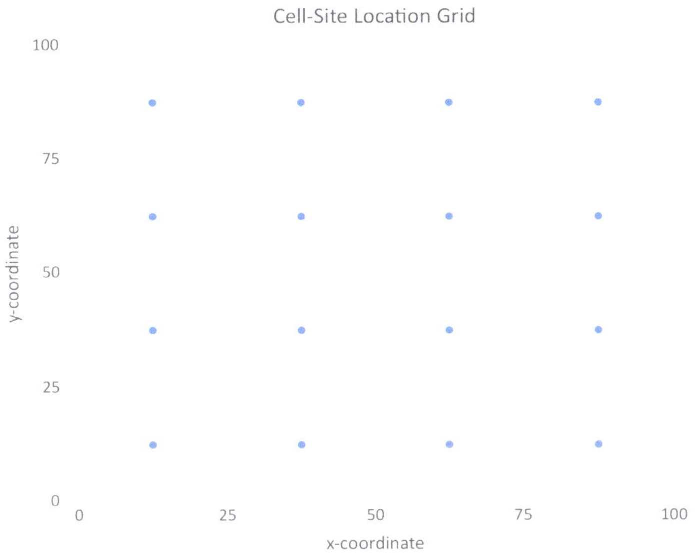

Figure 15: *Grid showing possible cell-site locations.*

Now it is time to place the four cell sites in our simple example. With four cell sites, you can define the location manually. However, the following code fragment will allow you the flexibility to define the locations for many more than four cell sites.

```
k = 1;
for i = 35:70:105
    for j = 35:70:105
         xbase(k) = xb(i,j);
         ybase(k) = yb(i,j);
        k = k + 1;
    end
end
```

This code generates the following xy-coordinates of the four cell sites:

Base station	x-coordinate	y-coordinate
1	862.5	862.5
2	2612.5	862.5
3	862.5	2612.5
4	2612.5	2612.5

Table 7: *Proposed cell-site locations.*

You may guess why I picked the cell-site locations by looking at the figure below.

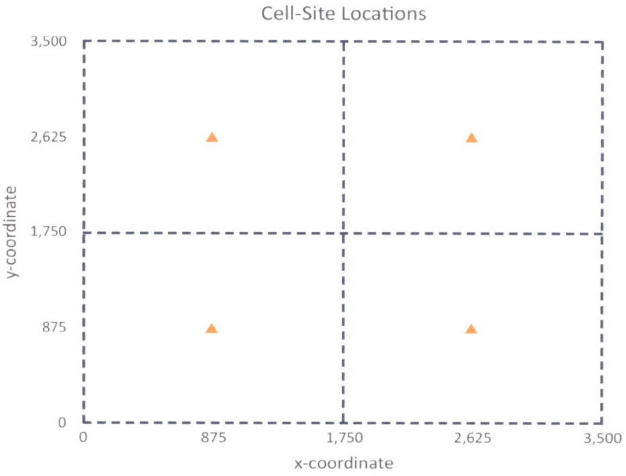

Figure 16: *Initial cell-site locations.*

As you can see, the study area was divided into four squares of equal size and the cell sites were placed in the middle of each square.

The next step is to select a predefined number of points in the mobile grid of Figure 14 to represent the mobile subscriber locations. The first thing we do is place all the possible mobile locations in a pseudo-column vector form:

```
xy = [reshape(xcoord,[],1), reshape(ycoord,[],1)];
```

I call it a pseudo-column vector because a column vector should have only one column, not two. If you interpret each row as a single entry, then `xy` can be considered as a column vector. The first few entries of the vector are shown in Table 8:

x-coordinate	y-coordinate
0	0
0	25
0	50
0	75
0	100
0	125
0	150
0	175

Table 8:
First few entries of `xy` *vector.* `xy` *is a vector with 19,881 entries.*

Now we will select a number of mobile subscriber locations with the following code:

```
sublocxy = xy(randperm(size(xy,1),numloc),:);
```

The number of locations is specified by the variable `numloc`. It's a bit of an art to select the number of locations to get meaningful results. If the number is too small, then vastly different results will be produced when the program is executed with different mobile locations of the same number (i.e. a different snapshot). When the number is too big, program execution time may become a problem, but a more important reason will be given below when we examine the effects of the geographical distribution of subscriber locations on network KPIs. Typically, I aim to have approximately 20 mobile locations for each cell sector. With this simple four-cell site example, 80 mobile subscriber locations will produce results that do not vary much for different snapshots. Table 9 shows the first few entries of `sublocxy`.

x-coordinate	y-coordinate
1225	2050
2925	475
2050	1800
1925	1750
3225	1050
1000	1025
2650	2575
2650	825

Table 9: *First few entries for mobile subscriber locations found in* `sublocxy`.

Mobile-subscriber locations along with the cell-site locations are shown in the figure below.

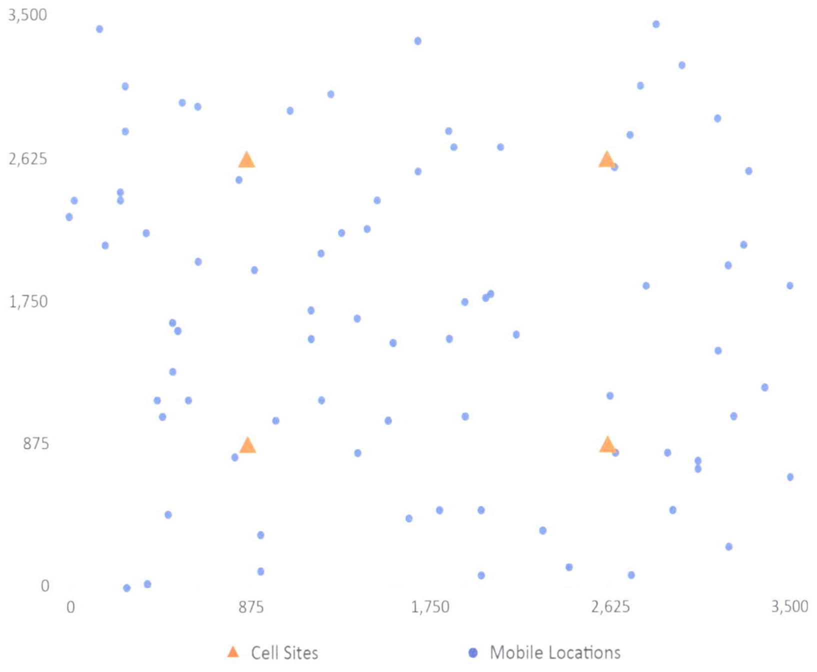

Figure 17: *Mobile-subscriber and cell-site locations.*

Although the mobile-subscriber locations appear random, when you run the MATLAB program over and over it will produce the exact same random locations if you use the following command at the start of the program: `rng('default')`. The performance of the network depends on the mobile-subscriber locations, so if the mobile-subscriber locations change from one simulation to the next, you won't know if changes to the KPIs are due to altered network parameters you are testing or the location changes of the mobile subscribers. Hopefully, the network parameter being adjusted will have a far greater impact on network performance than the exact locations of the mobile subscribers. But if the locations of the mobile subscribers are exactly the same between two tests, then we can rule out the possibility that changes to KPIs are due to altered mobile locations. In a sense, I'm elevating the status of the subscriber locations by using the same ones for different simulations. The network analysis approach becomes more user-centric. One can determine if a sufficient number of mobile-subscriber locations are selected by running the test a number of times, with the same number of locations but with different coordinates. This can be done by using `rng('shuffle')` before randomly selecting the mobile-subscriber locations. If the KPIs of differing snapshots are significantly different, then more mobile-subscriber locations should be selected. From experience, I found that when the average number of mobile-subscriber locations per radio is approximately 20, network performance is largely independent of the exact location of the mobile subscriber.

As mentioned above, there is another reason for not selecting too many subscriber locations per cell-site sector, besides limiting program running time. There is a network key performance indicator called the fifth (or tenth) percentile cell-edge user-perceived throughput which measures the slowest UPT in the network. Recalling that we assume that the user-perceived throughput of all subscriber locations within a sector are identical, the mobile-subscriber locations with the lowest UPTs are most likely residing within the same sector. Referring to Equation (11), the network parameter that affects the UPT the most is the sector utilization. Sector utilization is mostly affected by the number of subscribers. By having an average of only 20 subscriber locations per sector, it is likely some sectors will have many more subscriber locations than other sectors. This will result in some sectors becoming congested sooner than others, which is representative of an actual network. Cell sectors that congest soonest (i.e. sectors with the most subscribers) regulate the cell-edge UPT.

You may have noticed I've been using the term "subscriber location" over and over. This is to resolve a potential problem. In this section, I limit the average number of randomly placed subscriber locations to approximately 20 per sector so that I can take advantage of the Poisson process to vary the number of subscriber locations per sector. However, in the previous chapter, I mentioned that the number of active subscribers per sector should be on the order of several dozens for the user-perceived throughput Equation (10) or equivalent (11) to be valid. To resolve this problem, I allocated a number of subscribers at each subscriber location. The exact number of subscribers at each subscriber location is not important as long as the number per sector is at least on the order of several dozen.

Once the mobile subscriber and cell-site locations are fixed, the distance between each subscriber location and each cell site can be calculated:

```
distsub(:,i) = sqrt((xbase(i)-sublocxy(:,1)).^2+(ybase(i)-sublocxy(:,2)).^2);
```

The rows of `distsub` represent the distance between a particular subscriber location and each of the cell sites. The columns of `distsub` represent the distance between a particular cell site and each of the subscriber locations.

distsub: distance between sub location and cell site		Cell Site			
		1	2	3	4
Subscriber Location	1	1242	1826	669	1497
	2	2099	498	2970	2160
	3	1513	1093	1439	988
	4	1384	1123	1369	1103
	5	2370	641	2832	1678
	6	213	1621	1593	2263

Table 10: First few entries for `distsub`.

The main reason for calculating these distances is to determine the RF path loss between each cell site and each mobile subscriber. We will use a variant of the COST Hata RF path-loss model (31) mainly because it is used by 3GPP. It has many variables, but the most important one is the distance (d) between the cell site and the mobile subscriber.

The main reason for calculating the path loss is to determine the signal-to-interference-plus-noise ratio, which will eventually determine the possible data-link speed. The signal-to-interference-plus-noise ratio is simply defined as:

$$SINR = \frac{P}{I + N} \tag{35}$$

Where P is the power of the desired signal, I is the sum of the power of all the interfering signals and N is the power of thermal noise.

There are two other ratios very similar to SINR that are of interest, because they provide insight into mechanisms that affects the performance of wireless networks. The first is the carrier-to-interference ratio (*C/I*). *C/I* is defined as the power of the desired signal over the sum of the power of all the interfering signals, i.e. *P/I* in Equation (35). The other is the SNR, defined as the power of the desired signal over the power of thermal noise, or *P/N* in Equation (35).

You will notice throughout this book that I use each of the three versions of signal-to-noise ratios in different situations. Highlighting these three versions makes my approach different than other simulations of wireless networks, since they primarily focus on the most general form of SINR exemplified by Equation (35), as this formula is required to determine the data speed of the RF link. My method is not just about simulating wireless networks, but to unearth insights from simulations.

Rewriting Equation (35):

$$SINR = \frac{10^{Desired\ Power\ (dB)/10}}{\sum 10^{Interfering\ Power\ (dB)/10} + 10^{rnf/10}} \tag{36}$$

Where rnf is the receiver noise floor in dBm. When working with power, you need to define a bandwidth. With LTE, the typical choices for bandwidth are either the bandwidth of the Resource Element (15 kHz) or one of the supported LTE bandwidths (1.3, 3, 5, 10, 15, 20 MHz). Since I'm using the data speeds vs SNR chart of Figure 4, which is based on a 10 MHz carrier, I've decided to use 10 MHz as the bandwidth for all power calculations. When calculating the receiver noise floor, you must also include the receiver noise figure of the user equipment, which is typically around 7 dB.

$$rnf = 10 \log kTB + noise\ figure,$$
$$rnf = -174 + 70 + 7 = -97\ dBm.$$

In calculating the Desired Power (dB) and the Interfering Power (dB) in Equation (36), we need to use (34). The resulting SINR equation will be:

$$SINR = \frac{10^{(Transmit\ Power_i + Gains_i - Path\ Loss_i - Misc\ Loss_i)/10}}{\sum_{j \neq i} 10^{(Transmit\ Power_j + Gains_j - Path\ Loss_j - Misc\ Loss_j)/10} + 10^{rnf/10}} \tag{37}$$

Note that subscript i is used to designate the link between the desired base station and the mobile subscriber, while subscript j is used to designate the links from all the interfering base stations and the mobile subscriber. There is nothing wrong in using (37) as it is written. However, a labour-saving device will allow the user to analyze an infinite number of scenarios with each simulation. The problem with using (37) directly is that it is valid for one particular scenario. For example, when calculating the path loss, it is likely that Equation (31) will be used. With Equation (31), you need to specify the frequency of the carrier, the height of the base station, and the height of the mobile. If you change one of those parameters, the path loss changes and you have to redo the simulation with those changes. Or do you? The trick is to separate the factors that are unique to each subscriber locations from factors that are common amongst them. Interestingly, the unique factors affect *C/I* while the common factors affect SNR.

3.2 Capacity Analysis

The two most important KPIs of a wireless network are its capacity and its coverage. I will later show that the capacity of the network is limited by *C/I* while the coverage is limited by SNR. With the simple example we are examining, distance—or more specifically, relative distance—is key in calculating *C/I*.[25]

First, we eliminate the term associated with thermal noise in Equation (37):

$$C/I = \frac{10^{(Transmit\ Power_i + Gains_i - Path\ Loss_i - Misc\ Loss_i)/10}}{\sum_{j \neq i} 10^{(Transmit\ Power_j + Gains_j - Path\ Loss_j - Misc\ Loss_j)/10}} \tag{38}$$

Since we are assuming that base stations are transmitting at the same power, have the same omni-directional antennas with the same gain, the same amount of Misc. Loss and that all user equipment has the same antenna gain, we can rewrite (38) as:

$$C/I = \frac{10^{-Path\ Loss_i/10}}{\sum_{j \neq i} 10^{-Path\ Loss_j/10}} \tag{39}$$

We have a choice in path-loss equations. (See section 2.5.) We shall use an equation that combines Equation (31) and Equation (32).

$$PL = 10\gamma \log\left(\frac{d}{1000}\right) + 45.5 + (35.46 - 1.1h_m)\log f_c - 13.82 \log h_b + 0.7h_m + C + F. \quad [dB] \tag{40}$$

The assumption is that field measurements are used to determine γ, and the other constants shown above are illustrative of those if measurements are used in deriving Equation (40). In the absence of field measurements, one could either use the two-ray ground reflection model where the path-loss exponent is set to 4 or treat γ as a parameter

25 In the simple example we are using, both the base station and the mobile subscriber are using omni-directional antennas. If directional antennas are used for either base stations or mobile devices, the antenna patterns will also be factors in determining *C/I*.

to be adjusted to determine its effects on network KPIs.

To simplify the analysis, we make the assumption that all base-station antenna heights are the same and all mobile antenna heights are the same. Of course, the carrier frequency is the same for all cell sites and the correction factor *C* will be identical, since we are analyzing a limited geographical service area. Many factors in Equation (40) cancel and we get a simplified *C/I* equation:

$$C/I = \frac{10^{-\left(10\gamma\log\left(\frac{d_i}{1000}\right)+F_i\right)/10}}{\sum_{j\neq i} 10^{-\left(10\gamma\log\left(\frac{d_j}{1000}\right)+F_j\right)/10}} \tag{41}$$

Note that contained in the exponents of the terms in the numerator (desired signal) and denominator (interferers) are the loss components of the median path-loss equation attributed to *d*, which are unique for each subscriber location. (See Equation (33).) To reiterate, it is the unique component of the median path-loss equation related to *d* that affects *C/I*.

I'm going to make one more simplification, just to prove the earlier point about not worrying about the units of measurement when I defined the grid for mobile subscriber and cell-site locations. Assume that the log-normal fading component *F* is zero for all links. We have arrived at the simplest form of the *C/I* equation:

$$C\big/I = \frac{1}{\sum_{j\neq i}\left(d_i\big/d_j\right)^{\gamma}} \tag{42}$$

The *C/I* is only dependent on the ratio of distances (d_i/d_j). If we expand the distances by a certain scale factor, the *C/I* doesn't change. Also note that *C/I* increases when the path-loss exponent γ increases. In the above equation, the higher γ becomes, the smaller $(d_i/d_j)^\gamma$ becomes, making the denominator smaller and thus increasing *C/I*. If the wireless network is co-channel limited (i.e. thermal noise does not play a significant role), SINR matches *C/I*. If one replaces SINR with *C/I* and uses the Shannon-Hartley equation (1), then one can see that data-link speeds increase with increasing *C/I*.

Equation (42) is for explanatory purposes only, since it doesn't include the log-normal fading component. In fact, one could argue that *C/I* is only useful for explanatory purposes since what really matters is the SINR. When the distances between the subscriber locations and the cell sites are large, the impact of the thermal noise floor cannot be ignored, and we cannot use *C/I* to calculate data-link speeds. It will be shown later that for a particular geometry of base station and mobile-subscriber locations, *C/I* will determine the maximum capacity of the network.

Now let's revisit Equation (37). Exploiting the assumptions used in deriving Equation (39), we can rewrite Equation (37) as follows:

$$SINR = \frac{10^{-Path\ Loss_i/10}}{\sum_{j \neq i} 10^{-Path\ Loss_j/10} + 10^{(rnf - Transmit\ Power - Gain + Misc\ Loss)/10}} \quad (43)$$

Continuing with the same logic used in deriving Equation (41):

$$SINR = \frac{10^{-\left(10\gamma \log\left(\frac{d_i}{1000}\right) - 13.82 \log h_b + F_i\right)/10}}{\sum_{j \neq i} 10^{-\left(10\gamma \log\left(\frac{d_j}{1000}\right) - 13.82 \log h_b + F_j\right)/10} + 10^{(rnf - Transmit\ Power - Gain + Misc\ Loss + 45.5 + (35.46 - 1.1 h_m) \log f_c + 0.7 h_m + C)/10}} \quad (44)$$

Finally, recall that *C/I* is only dependent on the relative distances between the mobile subscriber and cell sites, but SINR is dependent on absolute distances due to thermal-noise considerations. Using this line of thought, I'm going to re-write the above equation:

$$SINR = \frac{10^{-\frac{\left(10\gamma \log\left(\frac{\tau d_i}{1000}\right) - 13.82 \log h_b + F_i\right)}{10}}}{\sum_{j \neq i} 10^{-\frac{\left(10\gamma \log\left(\frac{\tau d_j}{1000}\right) - 13.82 \log h_b + F_j\right)}{10}} + 10^{(rnf - Transmit\ Power - Gain + Misc\ Loss + 45.5 + (35.46 - 1.1 h_m) \log f_c + 0.7 h_m + C)/10}} \quad (45)$$

Where τ is a scale factor that we can use to change the distances between cell sites. If the path-loss exponent is the same for every cell site, we can rearrange the above equation to get the following:

$$SINR = \frac{10^{-\frac{\left(10\gamma \log\left(\frac{d_i}{1000}\right) - 13.82 \log h_b + F_i\right)}{10}}}{\sum_{j \neq i} 10^{-\frac{\left(10\gamma \log\left(\frac{d_j}{1000}\right) - 13.82 \log h_b + F_j\right)}{10}} + 10^{(rnf - Transmit\ Power - Gain + Misc\ Loss + 45.5 + (35.46 - 1.1 h_m) \log f_c + 0.7 h_m + C + 10\gamma \log \tau)/10}} \quad (46)$$

The last term of the above equation acts as a sort of adjusted thermal noise floor. We will give this term a symbol: η. Therefore:

$$SINR = \frac{10^{-\left(10\gamma \log\left(\frac{d_i}{1000}\right) - 13.82 \log h_b + F_i\right)/10}}{\sum_{j \neq i} 10^{-\left(10\gamma \log\left(\frac{d_j}{1000}\right) - 13.82 \log h_b + F_j\right)/10} + \eta} \quad (47)$$

η is the same for every subscriber location, much like the receiver noise floor 'rnf' it is designed to replace. When we develop the simulator, η becomes a parameter that is adjusted to determine its effects on network KPIs. If the base-station antenna heights are identical, the term $-13.82 \log h_b$ can be incorporated into η as it is the same for every RF link.

However, we shall keep the term separate as shown above, for reasons that will be explained later.

Note the similarity between Equation (47) and Equation (36). Some people may prefer Equation (36) due to its familiarity, but in the real world—where clients ask what the effects will be of changing various parameters such as transmit power, antenna height, in-building penetration loss, cell-site separation, frequency bands, carrier bandwidths, receiver-noise figure, etc.— you will appreciate Equation (47). Also note that *C/I* is a function of the geometry of the cell-site and mobile-subscriber locations. Once the relative locations of cell sites and mobiles are fixed on our grid (see Figure 17) and the log-normal fading components F_i are set, the *C/I* won't change unless you change the antenna pattern or path-loss exponent γ. You can change the scale of the x and y axis by modifying τ and the *C/I* won't change. Due to the added term η in the denominator, SINR can never be larger than *C/I*. To maximize SINR, we need to make η as small as possible. This can be done by:

- reducing the receive noise floor
- increasing the cell-site radio-transmit power
- increasing the antenna gain of the cell site/mobile
- decreasing the Misc. Loss (decreasing in-building losses, for example)
- lowering the spectrum band
- increasing the mobile antenna height
- decreasing the scale factor (i.e. moving everything closer together).

Although cellular engineers complain about co-channel interference, a worse situation is when the wireless network is not strictly limited by co-channel, but has issues related to thermal noise, i.e. high η. The main job of cell-site planners is to select cell-site locations and antenna patterns so that areas of high traffic also have high *C/I*. The upper limit of the performance of the wireless network will be determined by *C/I*. Once thermal noise η is introduced, the performance of the wireless network can only decrease.

From a network analysis perspective, a network that is strictly co-channel interference limited is far easier to analyze than a network where η plays a role. There is no need to evaluate coverage. If the interfering signals are much higher than thermal noise, the desired signal which is stronger than any interfering signal will also be much higher than thermal noise, thus, 100% coverage is assured. The spectral efficiency of the network can be expressed as a simple number and can be applied to any configuration of this network independent of spectrum band, transmitter power, receiver sensitivity or any other factor that can affect η.

For emphasis, here is the equation for η:

$$\eta = 10^{(rnf - Transmit\ Power - Gain + Misc\ Loss + 45.5 + (35.46 - 1.1h_m)\log f_c + 0.7h_m + C + 10\gamma \log \tau)/10} \quad (48)$$

Nothing in the above equation is complicated. It might be a bit of an administrative nightmare due to the number of terms, so this is where a spreadsheet is helpful.

As far as the model is concerned, a value for η is picked and its effect on SNR and SINR is observed (*C/I* is unaffected). The higher the value for η, the lower the SNR and SINR, the lower the data speeds and capacity.

Note that in rural areas, η is larger since the scale factor τ is larger. For a particular geometry of base stations and mobile subscriber locations, SINR will be lower in rural areas versus urban areas due to the larger value for η, which reside in the denominator of Equation (47) resulting in both lower data-link speeds and network capacity. However, quite often the spectral efficiency of busy cell sites in rural areas can be higher than for busy cell sites in urban areas. I do not doubt that busy rural cell sites can carry more traffic than busy urban cell sites, but the analysis in this chapter doesn't explain how this is possible. The heuristic method described in the first chapter uses this fact in the planning process. I also previously mentioned that I used OMs to confirm the validity of the stochastic method. More on this later.

The following code fragment generates the path loss between each cell site with all mobile locations.

```
for i = 1:numcells
    distsub(:,i) = sqrt((xbase(i)-sublocxy(:,1)).^2 + ...
        (ybase(i)-sublocxy(:,2)).^2);
    plosssub(:,i) = (10*rfexp)*log10(distsub(:,i)/1000) - ...
        13.82 * log10(hb);
end
cplosst = plosssub + lognormstd*randn(size(plosssub)); %path loss
with added lognormal fading F_i. Typically lognormstd is set to 10dB
```

Recall that Table 10 contains a matrix with the first few entries of `distsub`. Table 11 is a matrix with the first few entries of `cplosst`. Note that smaller numbers for `cplosst` (i.e. more negative numbers) indicate smaller path loss. Mobile subscribers will select the cell site with the smallest path loss. Therefore, for subscriber location 1, the mobile subscriber will select Cell Site 2 (`cplosst(1,2) = -30.17`) even though it is the furthest cell site away. (See Table 10.) For those who would rather use the COST Hata model Equation (31), the numbers in Table 11 are associated with a network where the base-station antenna height (hb) is 5.6 metres. This antenna height makes the path-loss exponent (γ = $(44.9\text{-}6.55\log_{10}h_b)/10$) equal a nice round number: 4. This is the same path-loss exponent used by the two-ray ground-reflection model. Also, 5.6 metres is the typical height of an omni-directional small cell in an urban environment.

Table 11: *First few entries of* `cplosst`.

`cplosst`: pathloss with log-normal fading (dB)		Cell Site 1	2	3	4
Subscriber Location	1	6.93	-30.17	-21.14	2.75
	2	0.29	-27.03	15.06	1.86
	3	-9.04	3.64	4.24	-3.55
	4	-7.63	-19.00	-15.04	-5.94
	5	-3.83	-8.74	3.04	3.60
	6	-48.42	1.55	-0.87	-10.98

It is interesting to observe the first few entries of `plosssub` in Table 12, which is the path loss without the log-normal fading term, or alternatively, the standard deviation of the log-normal fading term F_i is set to 0 dB. Much like the discussion pertaining to fixing the location of subscribers so that each simulation uses the same ones, when calculating the RF path loss for each base-station-to-subscriber-location link, F_i should not change between simulations either. There is no point in fixing the subscriber locations and not do the same for the log-normal fading term, as this will affect the subscriber location's SINR, data-link speed and serving base station. Hence, the table entries in Table 11 will not change from simulation to simulation if the same base-station locations, subscriber locations and F_is are used.

Table 12: *First few entries of* `plosssub`.

`plosssub`: pathloss without log-normal fading (dB)		Cell Site 1	2	3	4
Subscriber Location	1	-6.58	0.12	-17.32	-3.33
	2	2.54	-22.46	8.57	3.04
	3	-3.15	-8.79	-4.02	-10.54
	4	-4.69	-8.33	-4.89	-8.64
	5	4.65	-18.08	7.75	-1.34
	6	-37.21	-1.95	-2.25	3.85

Before we leave this section, I will introduce two MATLAB variables that will be used in future sections of this book; these indicate which cell site a mobile subscriber is attached to and how many subscribers are connected to each cell site.

```
[sploss,ixsubt] = sort(cplosst,2);
ixsub = ixsubt(:,1);
```

The above MATLAB statement sorts each row of `cplosst` into `sploss`, with the smallest number in the first column and the largest number in the last column. Interestingly, we never need to use `sploss` anywhere. The more important variable is `ixsubt`, which tells you from which column each entry in `sploss` originated.

`ixsubt`: index of sorted path loss		Cell Site			
		1	2	3	4
Subscriber Location	1	2	3	4	1
	2	2	1	4	3
	3	1	4	3	2
	4	2	3	1	4
	5	2	1	3	4
	6	1	4	3	2

Table 13: *First few entries of* `ixsubt`.

Only the first column of `ixsubt` is of interest, which tells you to which cell site each mobile subscriber is attached. This column is copied into `ixsub`.

This next variable tells you how many mobile subscribers are attached to each cell site.

```
subload = countmember((1:numcells)',ixsub);
```

The subroutine `countmember` is not a function provided by MATLAB. It was written by Jos ver der Geest and was downloaded for free from the Internet. Below is the description of `countmember`.

```
function C = countmember(A,B)
% COUNTMEMBER - count members
%
%   C = COUNTMEMBER(A,B) counts the number of times the elements of array A are
%   present in array B, so that C(k) equals the number of occurences of
%   A(k) in B. A may contain non-unique elements. C will have the same size as A.
%   A and B should be of the same type, and can be cell array of strings.
%
%   Examples:
%     countmember([1 2 1 3],[1 2 2 2 2])
%        -> 1     4     1     0
%     countmember({'a','b','c'},{'a','x','a'})
%        -> 2     0     0
%
```

Although 3GPP spends much effort simulating networks where all the radios are constantly transmitting data, radios in actual networks only transmit data a fraction of the time. Many air-interface standards specify that the radio constantly transmit a pilot signal, as in the case of LTE where the Cell-specific Reference Signal (CRS) is transmitted in every downlink subframe. (See section 2.2.) In the absence of data, LTE transmits the Almost Blank Subframe (ABS) with the CRS and other control signals preventing the subframe from being totally blank. To model the impact of the ABS, the equation is modified by adding a term in the path-loss term of the interferers. (See Equation (47).)

$$SINR = \frac{10^{-\frac{\left(10\gamma \log\left(\frac{d_i}{1000}\right) - 13.82 \log h_b + F_i\right)}{10}}}{\sum_{j \neq i} 10^{-\frac{\left(10\gamma \log\left(\frac{d_j}{1000}\right) - 13.82 \log h_b + F_j + ABS_j\right)}{10}} + \eta} \quad (49)$$

In the case of LTE, ABS_j is set to 0 dB in the above equation when the interfering base station is transmitting data and 10 dB when the interfering base station is transmitting the ABS. (See section 2.2.)

3.3 Coverage Analysis

A wireless network is not exclusively co-channel interference limited when η is large enough to affect the SINR and the data speeds of the RF links. When η is sufficiently large, the receiver may not be able to demodulate the desired signal and is deemed to be out of coverage. To assess the coverage of a wireless network, we need to establish the weakest signal strength that the receiver can demodulate (termed the minimum-required signal strength) which is determined by the minimum acceptable signal-to-noise ratio. This value is different for different air-interface standards. LTE can demodulate signals as low as -7 dB SNR[26], while Wi-Fi requires SNRs of at least 4 dB. There is also a trade-off between coverage and capacity, which will be examined later in this book. Accommodating lower SNRs means there are links with lower data speeds, thus lowering the capacity of the network.

We can modify either Equation (36) or (47) by eliminating the co-channel interference term.

$$SNR = \frac{10^{Desired\ Power\ (dB)/10}}{10^{rnf/10}}. \quad (50)$$

$$SNR = \frac{10^{-\left(10\gamma \log\left(\frac{d_i}{1000}\right) - 13.82 \log h_b + F_i\right)/10}}{\eta} \quad (51)$$

I prefer Equation (51) for reasons I will show later. When using Equation (51), the received signal strength in dBs (found in the exponent of the numerator in Equation (51)) is the negative of the component of

26 H. Holma & A. Toskala, *WCDMA for UMTS: HSPA Evolution and LTE*, John Wiley & Sons, 2010.

the median path-loss Equation (33) related to distance (i.e. $10\gamma\log(d)$), with adjustments for log-normal fading F_i and, if necessary, the base-station height. Once a value for η and the lowest acceptable SNR are selected, we can determine how many points in the mobile user-location grid meet the criterion. This can be done by running a Monte Carlo simulation, or it can be calculated by using a simple formula. Unless you are proficient in probability and statistics, I suggest Monte Carlo.

First, define the minimum-required SNR (SNR_{min}) for mobile subscribers to be considered in coverage. Now, calculate the smallest signal that mobile subscribers can receive[27] by multiplying the minimum-required SNR by the receive noise floor (if you are using Equation (50)) or by η (if you are using Equation (51)). For example, if the minimum-required SNR is 1 (or 0 dB), then the minimum-required signal strength (minSS) would be equal to the value η.

When calculating coverage, you don't need to select a limited number of points based on the number of cell sectors. You can select all the points in the mobile location grid in Figure 14. Since you are doing this simulation only once, computation time is not an issue.

```
minSSdB = 10*log10(SNRmin * nu); %minimum signal strength requirement
mcplosst = max(-cplosst,[],2); %finding the strongest signal
covered = mcplosst(mcplosst>minSSdB); %covered if strongest signal>minSSdB
coveragemeasured = length(covered)/length(mcplosst); %percent of loc'n
covered
```

In the section where I describe the simulation part of the program, only those mobile subscribers who meet the minimum-required SNR are allowed to be included in the analysis. In some analyses, the minimum-required SNR could be 0 or -∞ dB. This means all mobile subscriber locations are included in the analysis, regardless of its SNR. The following code fragment shows how the MATLAB variable `ixsub`, described in the previous section, can be filtered such that only locations meeting the minimum-required SNR and minimum-required signal strength are included:

```
ixsub = ixsub(mcplosst > minSSdB);
```

If `SNRmin` is set to 1 and `nu` is set to 1, then `minSSdB = 10*log10(SNRmin * nu)` will be 0 dB. Referring to Table 11, all rows in `cplosst` have an entry less than 0 which means that `mcplosst = max(-cplosst,[],2)` will have an entry greater than 0 dB, so all subscriber locations in Table 11 meet the minimum-required SNR.

27 In section 2.7, the term "minimum-required signal strength" defines the smallest signal that mobile subscribers can receive.

If `SNRmin` is set to 1 and `nu` is set to 10, then `minSSdB = 10*log10(SNRmin * nu)` will be 10 dB. Referring to Table 11, rows corresponding to Subscriber Location 3 and 5 do not have entries less than -10 dB, hence `mcplosst = max(-cplosst,[],2)` will not meet the minimum-required signal strength as specified by `minSSdB`.

3.3.1 Downlink Uplink Imbalance

Although wireless traffic analysis is focused predominantly on the downlink, since it can carry >90% of the traffic, uplink limitations may affect which subscriber locations are considered to be in coverage. This can be a problem especially for TD-LTE, where it is common to allocate 75% of the Resource Blocks for the downlink, significantly impacting uplink speeds. TCP/IP can accommodate asymmetric downlink-uplink data speeds. However, if the downlink speeds exceed uplink speed by a factor of 20, downlink speeds may be affected due to lost or delayed TCP acknowledgements.[28]

The discussion in the previous section is unquestionably appropriate for unidirectional wireless communication links (for example over-the-air TV broadcast) or links between devices with similar transmit output power (for example two Wi-Fi devices communicating with each other). For many other wireless air-interface standards, the power of the radios can differ vastly. In the case of LTE and HSPA, the base-station radio is over 100 times more powerful than the smartphone transmitter. There are techniques that both LTE and HSPA can use to keep the maximum path loss for the downlink and uplink similar. For example, in the case of LTE, the device can be restricted to transmitting over a limited number of Resource Blocks. This will severely reduce the data speeds for the uplink, to the extent that the downlink usage is discouraged or problems occur such as the one associated with asymmetrical link speeds, as outlined above. Thus, the uplink may govern the minimum-required signal strength and the minimum-required SNR for the downlink. Although LTE can support SNRs as low as -7 dB on the downlink, to establish a reliable uplink the RF path loss between the base station and mobile device must be reduced, resulting in a downlink minimum-required SNR that can be as much as 20 dB higher (SNR > 13 dB) for the device to be in coverage.

3.4 Calculating Capacity

The discussion thus far has focussed primarily on the input parameters required to calculate network KPIs, which included a brief explanation

28 H. Balakrishnan, MIT LCS, V. N. Padmanabhan, Microsoft Research, G. Fairhurst, M. Sooriyabandara, "University of Aberdeen, U.K., TCP Performance Implications of Network Path Asymmetry", *IETF RFC 3449*, December 2002.

for the decision to select a limited number of fixed subscriber locations. The following comparison of a deterministic and stochastic method to calculate capacity will provide additional insight for that design choice. As mentioned in the introduction of this chapter, the main difficulty in using Equation (6) is the fact that data-link speeds are constantly changing depending on whether potentially interfering cell sites are transmitting data, thus, making the calculation of s_i intractable via a deterministic approach. However, when all base-station radios are transmitting 100% of the time, the SINRs and the associated s_i for all subscriber locations do not vary and Equation (6) can be simply used. It is easy to establish the RF path losses for each point in the mobile location grid (see Figure 14), calculate SINR and s_i, and then use Equation (6) to calculate the capacity of each of the cell sites. Below is some MATLAB code that calculates SINRs for all of the 19,881 points (141 × 141) in the mobile location grid for the full-buffer scenario.

```
eploss = 10.^(-cplosst/10); %convert path loss to signal strength (Table 11)
mploss = max(eploss,[],2);  %find strongest signal
seploss = sum(eploss,2);
sinr = mploss./(seploss - mploss + Noise);
```

A lookup table based on Figure 4 can be used to find s_i and Equation (6) can be used to calculate sector capacity ρ_c.

One-point worth noting is that Equation (6) assumes that the base-station scheduler downloads the same amount of data for every point. By prioritizing subscribers with good SINRs, base-station equipment vendors have designed proprietary scheduling algorithms that will support traffic levels that vastly exceed the amounts calculated by Equation (6). These gains occur when the radio buffer is full, a scenario where even the common round-robin scheduler will support traffic levels exceeding Equation (6). However, MNOs eschew operating any cell-site radio at 100% utilization, thus negating the benefits of scheduling. The scenarios we want examined assume that the amount of data each subscriber downloads is independent of the data-link speed, as this is what is observed under normal network conditions where average network utilization is far below 100%.

The above MATLAB code fragment utilized every point in the mobile location grid to calculate sector capacity. One possible design choice is to select a subset of the points. One reason for doing this is to emulate the uneven cell-site traffic loading typical of actual networks. (See section 3.1.) This scheme works if A) all of the selected points download the same amount of data regardless of which sector they reside in, B) sectors have dissimilar amounts of subscriber locations, and C) no sector is at 100%

utilization.[29] However, in the full-buffer scenario, it will not be possible for subscriber locations in one sector to download the same amount of data as those in another sector, since the number of subscriber locations and the capacity of each sector will differ. Hence, there is another reason to use a limited number of points: reduction in program execution time. The deterministic approach used by the above MATLAB code fragment is very simple and can be processed quickly even when every point in the mobile grid is used. Selecting a subset of fixed points is optional for the full-buffer scenario since the deterministic approach can be used, but when sector utilization is less than 100% only the stochastic approach will work. The number of subscriber locations become important, as execution time is an issue with the stochastic approach. The stochastic approach can also be used for the full-buffer scenario and it will produce the exact same results as Equation (6), although it will take much longer.

The stochastic approach to be described resembles a Monte Carlo simulation. Rather than providing proof of the appropriateness of Monte Carlo simulation, I will first describe the algorithm and then explain how s_i, ρ and ρ_c are obtained. The goal is to develop a calculator that allows us to use Equation (6) for non-full-buffer scenarios. I will use terms associated with Monte Carlo simulation, although the algorithm I will describe does not aspire to capture all of the information provided by other commercially available Monte Carlo simulators. People assume that knowledge of probability and stochastic processes are required to learn Monte Carlo simulation. You will see that my algorithm requires none of that knowledge.

3.4.1 Algorithm to Calculate Capacity

A simulation consists of a set number of trials or epochs (typically in the tens of thousands) where in each trial a number of subscriber locations are selected to receive data from their serving base station. Referring to Figure 17, data can be downloaded to 80 possible mobile subscriber locations.[30] For each simulation, we are going to attempt to download the same amount of data to each user. For each trial in the simulation, we select a handful of the 80 subscriber locations to receive data. Since each selected subscriber location experiences distinct SINRs, the users in those subscriber locations will download data at different data rates, as determined by either Equation (1) or by using Figure 4. Since each subscriber location will have its own average download data

29 The reasons for these three conditions will become evident after the algorithm used to calculate capacity is described in the following sections.

30 In certain scenarios, not every subscriber location will be eligible to receive data. Some subscriber locations may not meet the minimum-required SNR and will be excluded from the simulation.

speed, it should be obvious that users in the subscriber locations with a lower average data speed should be selected more often than users with higher data speeds if all subscriber locations are to download the same amount of data. There are several ways to accomplish this task. The one I use involves the least amount of math background and offers the most flexibility. For each trial, the proportion of sectors transmitting data defines the network utilization for that trial. For a given simulation, the network utilization will not be the same for each trial. The goal, however, is to achieve a certain average network utilization for the simulation.

At time 0, each subscriber location is allotted a set amount of data. To avoid generalities, let's say that each subscriber location must download 100 Mb (100,000,000 bits). Whenever a subscriber location is selected, we determine its SINR and lookup the data speed by using Figure 4. For example, assume the user and network are provisioned to use 2x2 MIMO and the SINR at the subscriber location is 20dB. He or she is going to download data at 32 Mbps, assuming that 10 MHz spectrum is used[31]. If that subscriber location started out with an allotment of 100 Mb, and the user downloaded 32 Mb in the 1 second he or she was assigned the channel, the subscriber location will only have 68 Mb left to be received. The next time the subscriber location is selected, perhaps the quality of the channel degraded, and the data rate is reduced to only 20 Mbps. Now the subscriber location will only have 48 Mb of data remaining to be received. (See figure below.) This process will continue until there are no more data bits to be

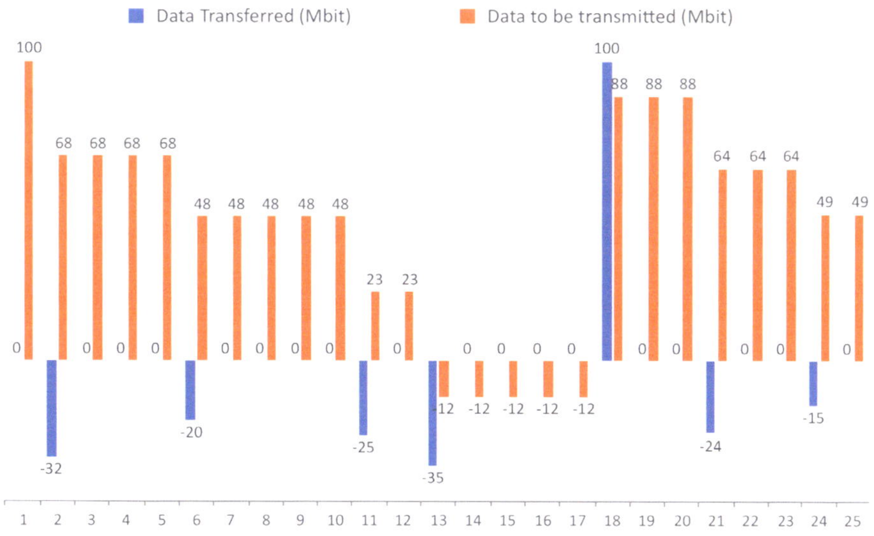

Figure 18: *Data account for a given subscriber location in a particular simulation.*

31 If the user is using 20 MHz of spectrum, the data rate will be 64 Mbps. If using 5 MHz of spectrum, the data rate will be 16 Mbps. In general, data rate = 32 * bandwidth (in MHz)/10.

received. Perhaps the last time the subscriber location was selected, more data was downloaded than was remaining in the allotment. That's okay; the remaining data will be some negative number. Now, that subscriber location will not be selected again until most subscriber locations receive at least 100 Mb. When most of the 80 subscriber locations have received at least 100 Mb, each subscriber location will be given another 100 Mb.[32] If the subscriber location downloaded more than 100 Mb the first time around, that location will have less than 100 Mb to download by the amount that exceeded 100 Mb. Or, in other words, if remaining data was some negative number, add 100 Mb to the negative number to get the new data allotment. Now all subscriber locations can be selected again, and the process continues as before.

A running total of the amount of data downloaded for each subscriber location is kept. Unless there are congested sectors, the goal is to have every subscriber location download the same amount of data.

For each trial or epoch, a subset of subscriber locations will have a positive balance of data and will be eligible to download data, while other locations will have already downloaded their allotment and will not be selected until receiving a sufficient amount of additional data during data top-up processes.[33] Of the subscriber locations with a positive balance of data, a certain number will be selected at random. Instead of downloading data to those selected subscriber locations, the cell sites to which these subscriber locations belong are marked as active, and the subscriber locations of each of those cell sites with the largest positive balance of data are chosen to receive data in its place. This will ensure that all subscriber locations in a cell site will download the same amount of data. If this substitution did not take place, it is possible that some SINR-disadvantaged subscriber locations will have an ever-growing positive balance of data, since they are always in the group of subscriber locations with unreceived data during the data top-up process.

How many subscriber locations should be selected during each trial? The more subscriber locations selected, the more data that is being downloaded, the higher the traffic being supported by the network.[34] The lowest number of subscriber locations to be selected is 0. No data is downloaded to any of the subscriber locations. Although this case appears trivial, not downloading any data in this epoch will affect the amount of traffic sent during the simulation.

The next lowest number is of course 1. Out of the 80 subscriber locations in the network, only one subscriber location is selected to

32 The 100 Mb top up amount is illustrative. I've used values between 1 Mb and 100 Mb for the data top up amount without observing any significant changes in the results.

33 A subscriber location may have such a large negative balance that it may require several data top ups before the balance becomes positive.

34 We will see later that this statement, while intuitive, is not always true.

download data. You should expect that the data rate this subscriber location is experiencing should be relatively high, as the SINR should be at its highest since no other base station will be transmitting data. With some wireless interface protocols, the denominator of the SINR equation (Equation (35) or (36)) consists of only the thermal noise term. With LTE, the interference term does not go to zero. Referring to section 2.2, LTE always transmits Cell-specific Reference Signals (CRS) even when there are no data to be transmitted. These signals contribute to the interference term in Equation (35) or (36), although to a lesser extent than when data are being transmitted (it is 10dB lower).

Also, most air-interface protocols also send information on control/signalling channels even when there are no data to be transmitted. The amount of resources consumed by the control/signalling channels can vary between 2% and 20% depending on the air-interface standard and the amount of control/signalling traffic required to be sent. In the case of LTE, it is possible to synchronize the time the control signal is sent from each cell site, so essentially the control-channel signal is not creating interference for subscriber downlink data. Synchronizing cell sites is a requirement for the TD variant of LTE (TD-LTE), while for the FD variant synchronization is optional and often not done. I added this discussion of the impact of control channel signalling for the sake of completeness, not because it will have a large impact on network capacity or UPT. In many cases, the data-link speed vs SINR graph may include the impact of control channel signalling. Perhaps some of the data can be attributed to control-channel signalling.

Getting back to the discussion of selecting subscriber locations, in the case of only selecting one subscriber location, there is no possibility of selecting more than one user from the same cell-site radio. When selecting more than one subscriber location, there is always the possibility that two or more subscriber locations will be served by the same cell-site radio. There are many ways to handle that situation. Some base-station vendors might claim it is possible to serve multiple users simultaneously by assigning each user a portion of the spectrum in a manner that maximizes data throughput. This is a technique called Frequency Division Scheduling. As you'll recall from the section "Capacity, Traffic and User-Perceived Throughput," it is rare that the network utilization averages more than 30%, so that in most circumstances scheduling does not occur. In the congested sectors where scheduling does occur, UPT is so degraded that the benefits of complex scheduler designs only make a bad situation marginally better. As described above, it does not matter which subscriber locations are selected, as the algorithm chooses the subscriber location for that cell site which has the most data in its buffer to be transmitted. If, in a given trial, two subscriber locations from the same sector are chosen to download data, the algorithm marks the cell site as active for the current as well as the next trial of the simulation, so that the sector will download data for consecutive trials. This will help the simulation achieve its targeted average utilization rate, especially when more subscriber locations are selected per trial, causing the network to approach congestion.

3.4.1.1 Determining the Number of Subscribers to be Selected per Epoch in a Simulation

It is useful to analyze how key performance indicators are affected by the level of network utilization. In each simulation, a certain utilization target is set. For each trial in the simulation, there is an attempt to meet the utilization target as closely as possible. If there are many sectors in the simulation, all one needs to do is select the number of subscribers per trial to establish the amount of network utilization. For example, if there are 100 sectors in the network, by selecting 10 subscriber locations per trial, approximately 10% of the network resources will be utilized. The reason that network utilization is not exactly 10% is because it is possible that some of the subscriber locations are being served by the same base-station radio. It may not be possible to achieve 10% network utilization even by sending data to subscriber locations that were possibly skipped in the previous epoch. In the end, it does not matter, since we will be running multiple simulations with different utilization targets ranging from low single digits to 100%. If it is not possible to achieve exactly 10% network utilization, we can estimate the KPIs for 10% network utilization by interpolating the indicators from simulations where network utilization is close to 10%.

If the network is sufficiently large (> 50 sectors), one method is to select the same number of subscriber locations at the start of each trial in the simulation. Detractors may argue that, in a real network, utilization will be constantly changing from one sample to the next. However, first, although I select the same number of subscriber locations at the start of each trial, the actual number of subscriber locations downloading data varies from one trial to the next due to:

- sectors with multiple subscriber locations selected in a trial
  - a sector can only download data to a single subscriber location per trial, thus some subscriber locations are excluded
- the inclusion of subscriber locations previously skipped in previous trials due to the previous point.

Second, if the network is large enough, different subsets of it will have a different number of selected subscriber locations for a given trial. Analyzing the number of selected subscriber locations for a subset of the network, the number of selected subscriber locations will change from trial to trial.

Unfortunately, in our example of the simple network with four omni-directional cell sites, we cannot use the simple strategy of selecting the same number of subscriber locations per trial. For example, to simulate a popular scenario of 25% network utilization, for each trial exactly one subscriber location should be selected. There will be no chance that there will be two subscriber locations selected in any trial, hence there will be no chance of co-channel interference. This is an unlikely scenario for a four-cell site network at 25% utilization. To solve this problem, I will use a simple technique. Of the 80 subscriber locations, select a subset of those locations where data is to be downloaded: for example, 2.5 times the number of sectors, or 10. (See Figure 19.) To simulate 25% network utilization, I still need to select, on average, one subscriber location uniformly distributed random number between 0 and 1. The number of

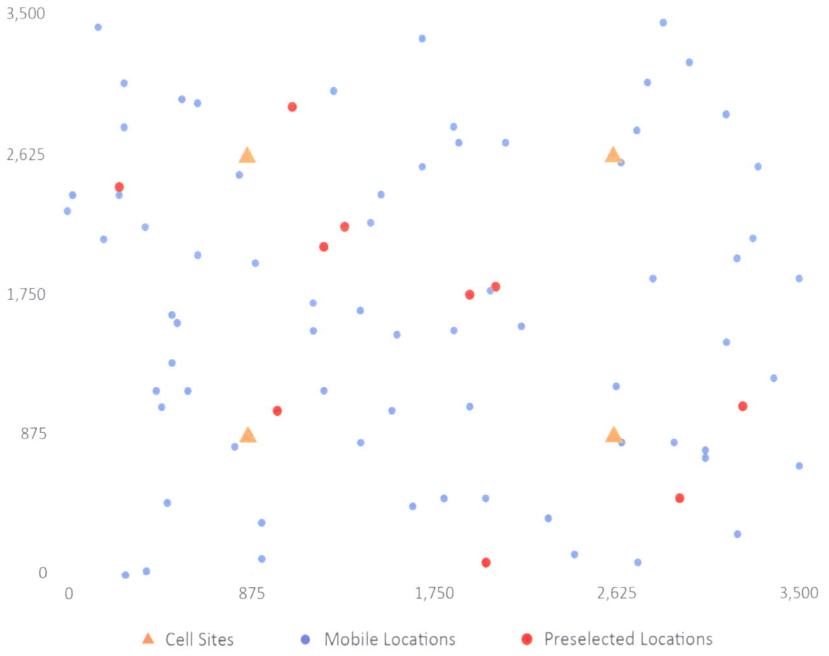

Figure 19: *An example where 10 of the 80 subscriber locations are preselected for possible data download.*[35]

35 A random number between 0 and 1 is generated for each of the preselected locations. If the random number falls below a value established to achieve the targeted utilization for the simulation, that location is selected to download data.

locations where the randomly generated number is less than 0.1 will be one, on average. However, it is possible to have more than one location where the randomly chosen number is less than 0.1, or no locations. Using this solution, the condition for topping up data is when there are fewer than 10 subscriber locations with data to be downloaded.

Another advantage of using this alternative method is that very low-traffic scenarios can be simulated. With large networks with dozens of sectors, selecting one subscriber per trial is a low-traffic scenario. With our simple network example, one subscriber per trial represents a utilization rate of 25%, which is typical of a busy network. Using this method, it is possible to have the average number of subscribers per trial be less than 1. For example, it is possible to have an average of only one subscriber being served in 10 trials. When a subscriber location is selected for download, it is likely to be the only location selected for that trial, hence there is very little chance of experiencing co-channel interference.

3.4.1.2 Running Sequential Simulations

I will often run a single simulation with a particular network utilization target. Most times, I perform multiple simulations on the same network with the identical subscriber locations with varying utilization targets. When I perform multiple simulations, I start with a low utilization target and then proceed to increase the utilization for subsequent simulations until I reach 100%. The reason for increasing the utilization target for sequential simulations will become apparent when we discuss the handling of congested sectors.

As more subscriber locations are selected per trial, you will find that the busiest sector is transmitting data in every trial of the simulation. If every subscriber in the network is downloading the same amount of data, then sectors that are not at 100% utilization will have unused resources that could deliver more data but can't, due to the busiest sector being congested. When resources of the busiest sector are fully consumed, it is impossible for that cell-site radio to serve subscriber locations in that sector the same amount of data as subscriber locations in other non-congested sectors. In that situation, subscriber locations in that sector should be segregated. Each subscriber location in congested sectors will receive the same amount of data, although it will be less data than subscriber locations in non-congested sectors receive. To determine if a cell is congested, measure how often that cell has data to transmit. If it is more than 99% of the time, it is considered congested.

As more subscriber locations are selected per trial, more sectors will become congested, until all sectors are congested. When all Resource

Blocks are carrying data, in most scenarios the wireless network will transmit the maximum amount of traffic.[36] However, the UPT will also reach its ultimate value of 0 Mbps. (Refer to section "Capacity, Traffic and User-Perceived Throughput" for an explanation of this).

While performing sequential simulations with increasing utilization targets, a sector that congests at a certain network utilization level will also congest when the utilization target is increased. Thus, at the start of a simulation, when it is known that a sector congests at a lower utilization level, the sector and its subscriber locations are segregated and treated separately from the rest of the network. In each epoch of the simulation, the subscriber location in the congested sector which has downloaded the least amount is then selected to receive data. Since this sector is always transmitting data, it will always be causing interference into neighbouring sectors. As traffic in neighbouring sectors increases, the capacity of the congested sector will decrease, as the subscriber locations within the congested sector will see greater co-channel interference.

When running sequential simulations, there is an adjustment that can be made after a sector is deemed congested. The previous section mentioned that 10 of the 80 subscriber locations were randomly chosen before running another test to determine which of the 10 will download data to meet the targeted utilization rate. When a sector becomes congested, the number of subscriber locations in non-congested sectors is reduced by the number of subscriber locations in the congested sector. To avoid large gaps in average network utilization rates and other potential problems, fewer than 10 subscriber locations from the non-congested sectors will be preselected before determining which will download data. Of the instances where a cell-site radio is transmitting data, we know the proportion that belongs to the congested sector. The number of subscriber locations to be preselected from non-congested sectors will be reduced by that amount. When a sector is deemed congested, let's say, for example, that the congested sector represented 35% of all the occurrences of a cell site radio downloading data. Once the congested sector is segregated, the number of subscriber locations that are preselected is reduced 35%, i.e. from 10 to 6.5. In all likelihood, that preselection number will not be an integer. This means that in every trial, either the largest integer less than the preselection number or the smallest integer greater than the preselection number will be used, such that the average number of subscriber locations preselected will equal the preselection number.

36 As mentioned previously, there are interesting situations where the maximum amount of traffic occurs when utilization is less than 100%.

For example, when the preselection number is 6.5, the number of subscriber locations preselected for each trial will be either 6 or 7, in equal amounts.

3.4.1.3 Sample MATLAB Code

Below is MATLAB code that implements the algorithm above. Instead of describing each variable of the program, I leave it as a challenge to the reader to examine each line of code and variable to determine how they relate to the algorithm. To increase readability, I did not include code to model the impact of control-channel signalling. I also excluded some code initializing memory for some matrices. MATLAB recommends you initialize space for matrices to speed up code execution. It is not a prerequisite to understand every line of code to read the subsequent material in this book. In fact, it might be a good idea only to peruse the code at first reading, even if your objective is to eventually run your own simulations.

```
subdatasum = zeros(numsubsr,1); % initialize subscriber download to 0
bucketsize = initialbucketsize; % typically set to 1-100 Mb
cellbank = zeros(numcells,1); % store requests for data download for a
sector
numdataspeedcell = ones(numcells,1); % number of times sector downloads data
congested = [];    % no cell sites are initially congested
effsubsr = 2.5*numcells; % select 10 sub loc'n as described in text
subsr = 0.1;        % select 10% of effsubsr
congest = 0.99;    % sector is congested if is busy more that 99% of time

for i = 1:trials  % trials is the number of times we select subs for
download

subbucket = bucketsize - subdatasum(1:numsubsr);
subque = find(subbucket > 0);
subqueue = subque(~ismember(ixsub(subque),congested));
noncongest = find(~ismember(ixsub,congested));
efftemp = floor(effsubsr+rand(1)); % adjustment for non-integer values
if subsr < 1 && length(congested) < numcells - 1 % not all cells congested
    if length(subqueue) > effsubsr
        rt = subqueue(randperm(length(subqueue),efftemp)); % select sub
index
    else
        if length(noncongest) <= effsubsr && effsubsr >= 1
            effsubsr = floor(length(noncongest)*0.8);
        end
        while length(subqueue) <= effsubsr && effsubsr > 0 % data top up
            bucketsize = bucketsize + bucketincrement; %bucketincrement
1-100
            subbucket = bucketsize - subdatasum(1:numsubsr);
```

```
            subque = find(subbucket > 0);
            subqueue = subque(~ismember(ixsub(subque),congested));
        end
        efftemp = floor(effsubsr+rand(1)); %adjustment for non-integer values
        rt = subqueue(randperm(length(subqueue),efftemp)); %select sub index
    end
    gatert = rand(effsubsr,1); %generate 10 uniformly random numbers from
0-1
    r = rt(gatert < subsr); %select subs whose random number < 0.1
    ixs = ixsub(r); %select subs to download data
    cellbank = cellbank+countmember((1:numcells)',ixs); %record cell
selected
    cellbank = min(subbank,10*ones(size(cellbank)));
    maskcell = ismember(1:numcells,ixs);
    subwithdraw = and(maskcell'==0,cellbank > 0);
    maskcell2 = or(subwithdraw',maskcell);
    cellactive = [congested; find(maskcell2)]; %these cells have data
    rsub2 = [];
    for j = 1:length(congested) %select subs from congested cells
        cell = congested(j);
        rsubo = find(ixsub==cell); %find all subs in the selected cell
        [sortsubbucket, isub] = sort(subbucket(rsubo),'descend');
        rsub2 = [rsub2 rsubo(isub(1))];%select sub with least data
downloaded
    end
    for j = 1:length(cellactivet)
        cell = cellactivet(j);
        rsubo = rsub(ixsub == cell); %find all subs in the selected cell
        [sortsubbucket, isub] = sort(subbucket(rsubo),'descend');
        rsub2 = [rsub2 rsubo(isub(1))];%select sub with least data
downloaded
        if (numdataspeedcell(cell)-1)/i > congest %test if cell is congested
            datatemp = numdataspeedcell(~ismember(1:numcells,congested));
            congested = [congested; cell]; %add cell to congested cell list
            numdataspeedcell(cell) = i;
            cellbank(cell) = 0;
            effsubsr = effsubsr*(1-i/sum(datatemp)); %select less subs/trial
        end
    end
    mask = ismember(1:numcells,ixsub(rsub2));
else %all cells are congested
    r = (1:numsubsr)';
    ixs = ixsub(r);
    cellactive = [];
    mask = zeros(1,numcells);
    rsub2 = [];
    for j = 1:numcells
        rsubo = r(ixs == j); %find all subs in the selected cell
        [sortsubdatasum, isub] = sort(subdatasum(rsubo));
        if ~isempty(isub)
            cellactive = [cellactive; j];
```

```
            mask(j) = 1;
            rsub2 = [rsub2 rsubo(isub(1))]; %select sub with least data
        end
    end
end
%place CoMP code here (see next chapter)
for j = 1:length(cellactive)
    cell = cellactive(j);
    cplosstemp = cplosst(rsub2(j),:); %get sub path loss see Table 11
    cplosstemp = cplosstemp + ABSoffset*ones(size(cplosstemp)).*(1-mask);
        %if cell site is not transmitting data, reduce power by 10dB/100dB
    eploss = 10.^(-cplosstemp./10);
    seploss = sum(eploss);
    sinr = eploss(cell)./(seploss - eploss(cell) + Noise);
    CtoI = eploss(cell)./(seploss - eploss(cell));
    StoN = eploss(cell)./(Noise);
    sinrdb = 10*log10(sinr);
    sinrdb = max(min(sinrdb,200),-40);
    sinrdbint = round(sinrdb)+11;%11dB offset used in data speed lookup
table
    sinrdbint(snrdbint > length(snrcurve)) = length(snrcurve);
    sinrdbint(sinrdbint < 1) = 1;
    dataspeedtemp = snrcurve(snrdbint) * TDDFDD; %get speed from Figure 4
        %store all data for processing later
    numdataspeedcell(cell) = numdataspeedcell(cell)+1; %increment sub index
    sinrsub(rsub2(j),numdataspeedsub(rsub2(j))) = sinrdb;
    CtoIsub(rsub2(j),numdataspeedsub(rsub2(j))) = 10*log10(CtoI);
    StoNsub(rsub2(j),numdataspeedsub(rsub2(j))) = 10*log10(StoN);
    dataspeedsub(rsub2(j),numdataspeedsub(rsub2(j))) = dataspeedtemp;
    subdatasum(rsub2(j)) = subdatasum(rsub2(j)) + dataspeedtemp;
    numdataspeedsub(rsub2(j)) = numdataspeedsub(rsub2(j)) + 1;
    subbank(rsub2(j)) = subbank(rsub2(j)) - 1; %indicate that sub is served
end
end
numdataspeedcell = numdataspeedcell-1; %adjustment for over counting by 1
numdataspeedsub = numdataspeedsub-1; %adjustment for over counting by 1
```

Some readers may find it more useful to write their own code without referring to the above fragment. This will be especially true for those who want to program the algorithm using another programming language.

The main point is that a useful simulation program does not require hundreds of thousands of lines of code. The above fragment is less than 200 lines. The shortness of the code fragment means that many aspects of the LTE standard are not being captured, for which I am unapologetic. The main purpose of this algorithm is to enable network dimensioning and the development of wireless strategies. The fact that I ignore numerous LTE protocols will not affect the comparison of various proposed features or network design alternatives to reduce costs.

3.4.2 Analyzing Subscriber KPIs

With the data now collected from the simulation, it is time to process the data. Looking at the above MATLAB code, the only data collected is related to the mobile subscriber: `sinrsub` (see Table 14), `CtoIsub` (see Table 15), `StoNsub` (see Table 16), `dataspeedsub`[37] (see Table 17), `subdatasum` (see Table 18), and `numdataspeedsub` (see Table 18). In this analysis of the simple four-cell-site network example, we set the adjusted thermal noise floor η to 10. Table 14 to Table 18 show data for a simulation that had a target of two subscriber locations selected per trial on average (50% sector utilization). We will discuss the implication of $\eta = 10$ later in the chapter. Also, the simulation had a minimum-required SNR (SNR_{min}) of 0 dB[38], which implies that the minimum-required signal strength is also 10. Of the 80 initial subscriber locations, only 69 met that criteria. Cell sites 1 and 3 both have 18 subscriber locations, while cell site 2 had 23 locations and cell site 4 had 10.[39]

`sinrsub`: Signal to Interference and Noise ratio (dB)		Sample Number			
		1	2	3	4
Subscriber Location	1	8.70	8.70	16.50	16.50
	2	16.96	16.38	16.61	16.61
	3	7.48	2.18	6.12	7.48
	4	37.43	37.43	37.60	37.22
	5	60.69	60.93	61.41	60.93
	6	49.35	49.34	49.39	48.39

Table 14: *First few entries of* `sinrsub` *(dB).*

`CtoIsub`: Carrier to Interference ratio (dB)		Sample Number			
		1	2	3	4
Subscriber Location	1	9.02	9.02	18.94	18.94
	2	34.94	24.94	27.01	27.01
	3	12.80	3.19	9.28	12.80
	4	44.35	44.35	45.24	43.40
	5	68.12	69.66	76.23	69.66
	6	63.19	62.98	64.31	54.76

Table 15: *First few entries of* `CtoIsub` *(dB).*

`StoNsub`: Signal to Thermal Noise ratio (dB)		Sample Number			
		1	2	3	4
Subscriber Location	1	20.17	20.17	20.17	20.17
	2	17.03	17.03	17.03	17.03
	3	9.00	9.00	9.00	9.00
	4	38.42	38.42	38.42	38.42
	5	61.56	61.56	61.56	61.56
	6	49.53	49.53	49.53	49.53

Table 16: *First few entries of* `StoNsub` *(dB). Note that entries for each subscriber location do not vary.*

37 All data-link speeds in this chapter assume that the base station and mobile receiver support 2×2 MIMO and 10 MHz of LTE is deployed. The 2×2 MIMO graph of Figure 4 is used to compute data-link speed as a function of SINR.
38 Desired signal strength should be greater than the receiver noise floor (i.e. thermal noise sources).
39 Note that the number of subscriber locations for each cell site of this wireless network is one-tenth the number of subscribers for each sink node of the wired network analyzed in the previous chapter. (See Table 4.)

Note that sinrsub is lower than either CtoIsub or StoNsub. When the difference between CtoIsub and StoNsub is greater than 10 dB, sinrsub follows the lower one fairly closely.

Table 17: *First few entries of* dataspeedsub *(Mbps), where two subscribers are receiving data per trial.*

dataspeedsub: data-link speed (Mbps)		Sample Number			
		1	2	3	4
Subscriber Location	1	11.80	11.80	26.70	26.70
	2	26.70	24.60	26.70	26.70
	3	9.40	4.60	8.20	9.40
	4	45.50	45.50	45.50	45.50
	5	45.50	45.50	45.50	45.50
	6	45.50	45.50	45.50	45.50

Table 18: *First few entries of* subdatasum *(Mbit) and numdataspeedsub, where two subscribers are receiving data per trial.*

		subdatasum	numdataspeedsub
Subscriber Location	1	2701.10	141
	2	2784.60	105
	3	2726.60	393
	4	2821.00	62
	5	2821.00	62
	6	2821.00	62

The algorithm tries to keep subdatasum the same for each subscriber, which means subscriber locations with lower data speeds get selected more often.

It is not often that you will be analyzing the KPIs for the subscriber locations, due to their large numbers. The subscriber-location indicators are aggregated at the cell-sector level and network level.

3.4.3 Analyzing Base-Station KPIs

This modeling approach focuses on the user, but some user KPIs, such as UPT, cannot be estimated without analyzing base-station KPIs. In the case of UPT, link utilization is required (see Equation (11)), which is a base-station KPI. Furthermore, people in Engineering and Operations focus on how well the cell sites are performing, since this is how they troubleshoot network problems, and it also affects network planning.

3.4.3.1 Sector Utilization

In LTE, sector utilization is also called Resource Block utilization, since all data are transported in Resource Blocks. In the main MATLAB simulation program, a variable called numdataspeedcell counted the number of times the base-station radio was transmitting data to a subscriber. It is also possible to use `sum(numdataspeedsub(ixsub==cell_number))` to arrive at the same number. Link or sector utilization is simply the number of times the base-station radio was transmitting data divided by the total number of trials in the simulation. Sector utilization can range from 0% (no data being transmitted) to 100%, when the base-station radio is transmitting data for every trial in the simulation.

In the section 2.3 titled "Capacity, Traffic and User-Perceived Throughput," utilization was given the symbol θ. The table below shows the sector utilization for the simple example we've been analyzing.

Sector Utilization	Targeted average number of selected subscriber locations per trial								
	0.5	1	1.5	2	2.5	3	3.5	4	4.5
Cell Site 1	0.12	0.25	0.37	0.49	0.59	0.67	0.90	1.00	1.00
Cell Site 2	0.18	0.36	0.54	0.71	0.85	0.96	1.00	1.00	1.00
Cell Site 3	0.11	0.23	0.34	0.44	0.53	0.58	0.76	1.00	1.00
Cell Site 4	0.08	0.16	0.24	0.31	0.36	0.40	0.54	0.97	1.00

Table 19: *Link utilization for different targeted average number of selected subscribers per trial.*

When a cell site reaches 100% utilization, that cell site is congested and UPT is theoretically zero by Equation (11).

3.4.3.2 Cell-Site Traffic and Sector Capacity

There are several ways to calculate cell-site traffic, typically measured in Mbps. Each method produces the exact same result but provides a slightly different perspective on the meaning of cell-site traffic. Once cell-site traffic (ρ) and sector utilization (θ) are determined, capacity can be calculated by using Equation (7), i.e. $\rho_c = \rho / \theta$.

The first method actually calculates sector capacity ρ_c by using Equation (6) first before using Equation (7) to determine cell-site traffic.

METHOD 1:

The first method involves finding the average data-link speed for each subscriber. In section 2.3 titled "Capacity, Traffic and User-Perceived Throughput," it was stated that when performing a time average of the data-link speed for an individual subscriber, an arithmetic average should be used. When calculating the capacity of the cell-site sector where each subscriber downloads the same amount of data, use the harmonic average.

Below is a modification of Table 18 with another column added, showing the average data-link speed for each subscriber location (`meansubspeed`).

		`meansubspeed`	`subdatasum`	`numdataspeedsub`
Subscriber Location	1	19.16	2701.10	141
	2	26.52	2784.60	105
	3	6.94	2726.60	393
	4	45.50	2821.00	62
	5	45.50	2821.00	62
	6	45.50	2821.00	62

Table 20: *First few entries of* `meansubspeed` *(Mbps),* `subdatasum` *(Mbit) and* `numdataspeedsub`*, where on average two subscribers are receiving data per trial (i.e. 50% network utilization).*

The MATLAB variable `meansubspeed` can be calculated in two ways. The first method involves calculating the average of each row in Table 17, i.e. the entries for `dataspeedsub`. Note that Table 17 only shows the first four samples of the data-link speeds `dataspeedsub` for each of the subscriber locations.

The second method comes from Table 20 directly.

(52)
```
meansubspeed = subdatasum / numdataspeedsub
```

Once the average data-link speeds `meansubspeed` have been calculated, we use the results from the section 2.3 to calculate the capacity of the cell site. We obtain `meansubspeed` for every subscriber attached to the cell site and use Equation (6) to obtain the sector capacity, noting that s_i = `meansubspeed`. Or by using MATLAB notation:

(53)
```
cell_capacity = 1/mean(1./meansubspeed)
```

The amount of traffic in the cell is obtained by using Equation (7), which is simply the cell capacity multiplied by the link or sector utilization ($\rho = \rho_c\theta$).

METHOD 2:

The second method is easier than the first. We know how much data the base station transmitted during the simulation by adding up the traffic (in Mb) each subscriber in the sector received. To convert traffic from Mb to Mbps, simply divide the traffic in Mb by the number of trials in the simulation. Sector capacity is calculated by using $\rho_c = \rho / \theta$.

Method 2 I feel is the more natural method for calculating sector capacity. Also, there is no need to calculate s_i for each subscriber location. Method 1 resembles the discussion of section 2.3. Both methods give the same results.

METHOD 3:

The average data-link speed (`meansubspeed`) is the average speed at which the user receives data when the base station is actively sending data to the user. (See METHOD 1.) The user is only receiving data for part of the time during the simulation. This percentage is equal to the total number of data speed samples collected during the simulation for a particular subscriber divided by the number of trials in the simulation. This basically measures how often the subscriber is using the link. Multiplying the average data-link speed by the utilization indicates how much traffic the subscriber is downloading in Mbps. If you add the traffic of all the users in the sector, you obtain the sector traffic in Mbps.

Yet another method is to convert the amount of traffic each subscriber received during the simulation and divide it by the number of trials in the simulation. This give the traffic a subscriber receives in Mbps. Once again, sum the individual subscriber traffic to obtain the sector traffic.

Sector capacity is calculated by using $\rho_c = \rho / \theta$.

3.4.3.3 Evidence of the Validity of the Algorithm

By providing so much detail of the algorithm, it is easy to not see the forest for the trees. At the start of section 3.4.1, the goal was to describe the algorithm that will allow us to calculate capacity using Equation (6). Equation (53) is the MATLAB version of Equation (6). So, have we accomplished what we set out to do without realizing it?

The fact that all three methods produce the same results signifies nothing, since they are all manipulating the same data. However, we know of an alternative method of calculating sector capacity by using Equation (6) when every sector is at 100% utilization. When we run the algorithm, something happens to `dataspeedsub` in Table 17 when all sectors are at congestion. All the entries for a given subscriber location become invariant and it matches the data-link speed used by the analytical/deterministic method. We could run the stochastic algorithm just enough times so that every subscriber location is selected at least once, knowing that in subsequent trials the subscriber location will be assigned the same number for `dataspeedsub`. At this point we could use Method 1 to calculate sector capacity, since `meansubspeed` will be the same as `dataspeedsub`. However, we cannot use the other methods (Method 2 & 3) because all of the subscriber locations will not have downloaded the same amount of data. It takes many trials until all of the subscriber locations have approximately the same amount of data. Looking at Table 20, all the subscriber locations have downloaded between 2,700 and 2,825 Mbits and all have been selected at least 62 times during the simulation. This explains why the deterministic method is so much faster, since it only needs to calculate the data-link speed once, whereas the stochastic method requires each subscriber location to be selected dozens of times before we can be confident that each location downloads the same amount of data even when SINR and s_i are constant due to 100% network Resource Block utilization. For the scenario where there is no transmit buffer congestion, each subscriber location must be selected multiple times, not only to ensure that each location downloads the same amount of data but to make sure that the average data-link speed for each location is not changing. The data-link speeds are changing from one trial to the next since co-channel interference is changing, causing the SINR to vary.

There is another scenario where the deterministic method can be used instead of the stochastic algorithm: when network traffic is extremely low. As in the scenario where all radios are at 100% Resource Block utilization, in the extremely low-traffic scenario, SINR also does not vary. When data is being downloaded to a particular subscriber, it is unlikely that another sector is transmitting data. In the case of LTE, only the CRS associated with the ABS frame is transmitted, which is 10 dB lower in power compared to a frame with data. Using the deterministic method, it is easy to calculate the SINR and the associated s_i for each subscriber location and then use Equation (6) to calculate sector capacity. The actual amount of traffic is arbitrarily low, based on the assumed amount of utilization. To ensure that only one radio is transmitting data, sector utilization θ should be lower than 1% even for the four cell-site toy network example. When the stochastic method is used, it is possible to set up the simulation where on average one subscriber per every 100 trials is selected to download data. Even with this extremely low network utilization, it is still possible for two cell sites to download data at the same time (this occurs approximately once every 30,000 trials). With the stochastic method, you can run simulations consisting of millions of trials in order for each subscriber location to be selected enough times so that all of them have downloaded the same amount of data at the end. The only purpose for running this scenario is to verify that both the deterministic and stochastic method produces the same results (which they do).

Note that when network traffic is extremely low, sector capacity is at its highest, as SINR and data-link speeds are at their highest. When network traffic is at its maximum, sector capacity is at its lowest and is equal to sector traffic. SINR and data-link speeds are also at their lowest.

The deterministic and stochastic methods give the same results at the bookends, at very low utilization and at 100% utilization. At all other levels of network utilization, only the stochastic method is available. The algorithm downloads the same amount of data to each subscriber location in the sector, each location has an average data-link speed s_i associated with it, and all the calculations for traffic and capacity described in Methods 1 to 3 agree, so the stochastic algorithm as a method to solve Equation (6) appears reasonable, although I wouldn't go as far as saying it has been rigorously proven. The lack of a rigorous proof should not be a reason not to use it, since it will provide substantial insights into the behaviour of wireless networks and the alternatives—

the application of heuristics while inspecting OMs and full-scale Monte Carlo simulations—do not appear attractive.

3.4.3.4 User-Perceived Throughput

There are two virtually identical equations ((10) and (11)) that can be used to calculate UPT. Note that both give the aggregate average user-perceived throughput for the sector. It is the average speed you would measure if you downloaded the same amount of data at each subscriber location in the sector divided by the total time it took to download all that data. It is not the equation for the UPT at a particular location. It was suggested that if we take the average data-link speed for a particular subscriber location and multiply it by $(1 - \theta)$, where the θ is the sector link utilization, we get the UPT for that subscriber location. Even though I advocate a user-centric approach for analyzing wireless networks, eventually we need to aggregate user results, since there are so many of them. In most analyses, we are comparing two network scenarios or looking for the approximate improvement of a new feature. So, as long as we are consistent in how we measure average UPT between different scenarios, the results of the analyses remain informative.

3.4.4 Analyzing Network KPIs

A network consists of a number of cell sites. With certain KPIs, such as network traffic, it doesn't matter whether you sum cell-sector traffic or sum subscriber traffic, since both totals will be the same. When calculating the average network UPT, it matters whether you use the cell-site UPT or the subscriber-location UPT. Maintaining the user-centric approach when performing network analysis, when there is a choice of calculating with cell site KPIs or with the equivalent subscriber KPI, I would choose subscriber KPIs. This is because there are large variations in the amount of traffic and subscriber locations each sector supports. We want the sectors carrying a disproportionate share of the traffic and subscriber locations to have a larger impact on the network KPI than on sectors carrying little traffic. The only exception is link utilization (or Resource Block utilization in LTE), although it is questionable if link utilization is a KPI. In any case, to obtain network utilization, simply sum the utilization of all cell sectors and divide by the number of cell sectors.

Network KPIs are similar to cell-site KPIs except for one additional network KPI called the cell-edge UPT. The cell-edge UPT is the lowest fifth or tenth percentile of UPTs within the network. In the discussion of UPT in the section analyzing cell-site KPIs, there are fewer problems using the aggregate UPT for the cell sector than using the UPT for each subscriber location. When calculating the network cell-edge UPT, each subscriber location in a cell sector is given the same aggregate UPT. Then the UPT from all the subscriber locations in the network are collected, sorted, and the fifth or tenth percentile is deemed the cell-edge. My characterization of the term cell-edge differs dramatically from established convention, which typically invokes images of some

Table 21: *Template for displaying network and cell-site KPIs.*

	Subs selected	0.05	0.1	0.15	0.2	0.25	0.279	0.3	0.35
Network KPIs	Trials	10000	10000	10000	10000	10000	10000	10000	10000
	Spectral efficiency (bps/Hz)	0.13	0.25	0.35	0.45	0.53	0.57	0.59	0.63
	Spec eff active RBs (bps/Hz)	1.03	0.99	0.95	0.90	0.86	0.83	0.82	0.79
	Traffic (Mbps)	5.16	9.86	14.17	18.06	21.24	22.80	23.72	25.28
	User-perceived throughput (Mbps)	8.96	7.32	5.75	4.24	2.98	2.28	1.95	1.35
	SINR (dB)	12.50	12.15	11.83	11.45	11.17	10.96	10.85	10.64
	C/I (dB)	23.55	22.09	20.74	19.42	18.33	17.70	17.32	16.70
	SNR (dB)	13.66	13.66	13.66	13.66	13.66	13.66	13.66	13.66
	10th cell-edge (Mbps)	7.79	5.82	4.00	2.18	0.79	0.00	0.00	0.00
	5th cell-edge (Mbps)	7.79	5.82	4.00	2.18	0.79	0.00	0.00	0.00
	utilization	0.13	0.25	0.37	0.50	0.62	0.69	0.73	0.80
Cell-Site KPIs	Traffic (Mbps)								
	1	1.35	2.57	3.70	4.71	5.54	5.98	6.43	7.23
	2	1.71	3.28	4.72	6.02	7.08	7.50	7.30	6.82
	3	1.35	2.58	3.70	4.72	5.54	5.98	6.43	7.22
	4	0.75	1.42	2.05	2.62	3.08	3.32	3.57	4.01
	User-perceived throughput (Mbps)								
	1	9.29	7.57	5.89	4.33	2.95	2.20	1.64	0.66
	2	7.79	5.82	4.00	2.18	0.79	0.00	0.00	0.00
	3	10.11	8.62	7.20	5.89	4.72	4.12	3.59	2.60
	4	9.00	7.98	6.94	5.86	4.92	4.33	4.04	3.45
	SINR (dB)								
	1	13.83	13.42	13.01	12.57	12.21	11.98	11.89	11.69
	2	11.35	10.94	10.62	10.14	9.89	9.62	9.45	9.17
	3	12.24	11.98	11.76	11.57	11.32	11.23	11.15	10.99
	4	13.26	12.94	12.60	12.21	11.94	11.73	11.67	11.50
	Utilization								
	1	0.13	0.25	0.39	0.52	0.65	0.73	0.80	0.92
	2	0.18	0.36	0.54	0.73	0.90	1.00	1.00	1.00
	3	0.12	0.23	0.34	0.44	0.54	0.59	0.64	0.74
	4	0.08	0.15	0.23	0.31	0.39	0.43	0.47	0.54

unfortunate subscriber in an area where there is no dominant cell-site server or stuck in an underground parking garage. It's not that I'm not concerned about those situations, but my definition also captures other scenarios that will cause low UPT.

In this first example of the simple four omni-directional cell sites, it is not too cumbersome to analyze the KPIs of each cell site. However, most realistic examples we look at have dozens, if not hundreds, of cell sectors. In the end, we are looking for some measure of network spectral efficiency in either bps/Hz/sector or GB/month/sector. This not only reduces the amount of data to be analyzed, but it also allows comparisons between networks with different numbers of cell sites/sectors.

It is possible that you will analyze dozens of scenarios, each requiring a separate simulation. To make this more efficient, use the same template to display the results of each simulation. See Table 21 for an example template for the simple four-cell site network we have been examining. The top part of the table shows the network KPIs, the bottom the cell-site KPIs.

To aid analysis, a number of graphs can be generated from the above table. The most important one is the Network User-Perceived Throughput vs Network Traffic graph. (See Figure 20.) This graph shows the trade-off between network traffic and network UPT. The more traffic the network carries, the slower the UPT. Note that Figure 20 shows both mean UPT and cell-edge UPT. Although it is possible for this network to support 30 Mbps of traffic, if the mobile network operator wants to have happy subscribers, perhaps only 20 Mbps is recommended if the quality of experience (QoE) of the fifth percentile cell-edge subscriber is important.

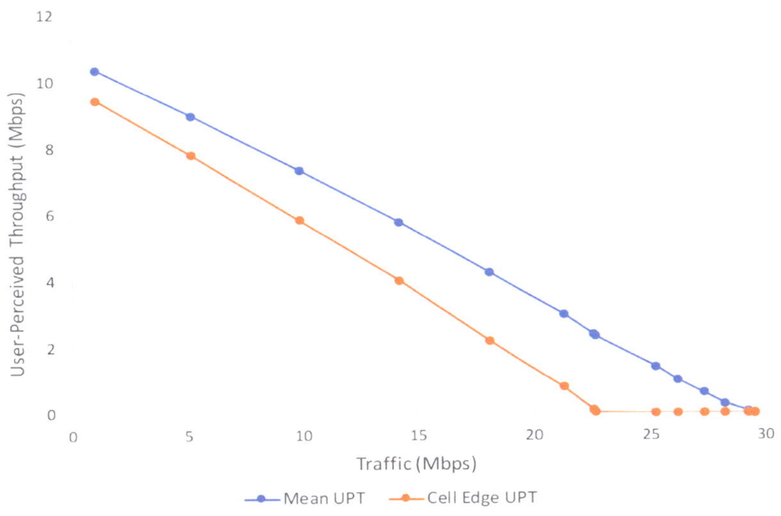

Figure 20: *Network user-perceived throughput vs network traffic.*

Note the similarities between the wireless network graph Figure 20 and the wireline network graph Figure 11. I purposely selected the number of subscribers at each end node of the wireline network to be 10 times the number of subscriber locations in each sector in the wireless network. Thus, the traffic distribution for the wireline network and wireless network are identical.

The figure below shows that the slope of the wireless network is slightly more negative than the wireline network.

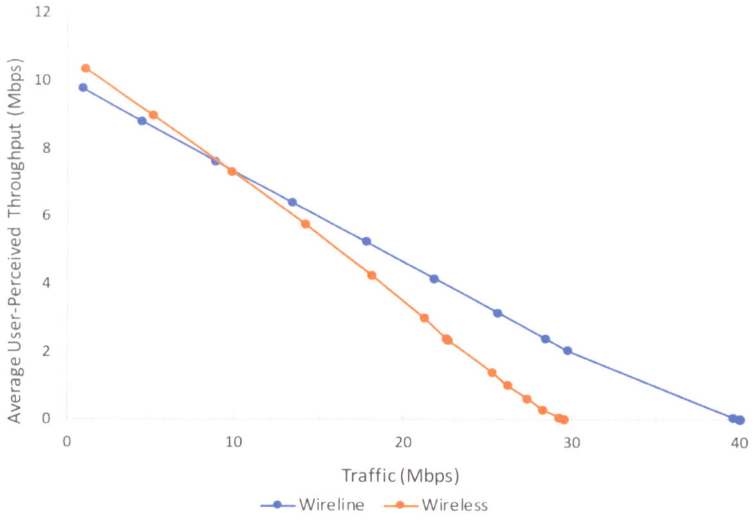

Figure 21: *Average UPT vs traffic for wireless and wireline networks.*

The reason for the slightly steeper graph for the wireless network can be found by analyzing signal-to-noise ratios graphs. (See Figure 22.) In the wireline network, data-link speeds do not change with traffic. In the wireless network, as traffic increases, SINR decreases due to increasing co-channel interference, causing data-link speeds to decrease. Later we will examine a scenario where co-channel interference does not change with traffic. It comes with a cost, and thus is shown for educational purposes only.

Under low-traffic loading, the SNR is much lower than the *C/I*, i.e., co-channel interference is much less detrimental than thermal noise. Hence, SINR is mostly determined by the SNR. As traffic increases, *C/I* becomes more significant, causing the SINR to decrease. Note that SNR is constant (i.e. independent of traffic) since it only measures the ratio of the desired signal and the thermal noise floor. Also, note that *C/I* is 10 dB lower when all base-station radios are transmitting data, compared to when the only co-channel interference comes from the cell-specific reference signals (when traffic and utilization are both 0).

The following graph (see Figure 23) shows the impact of increasing traffic and co-channel interference on the spectral efficiency of actively transported Resource Blocks. As traffic increases, the spectral efficiency of the network increases,[40] but the spectral efficiency of the active Resource Blocks decreases due to increased co-channel interference.

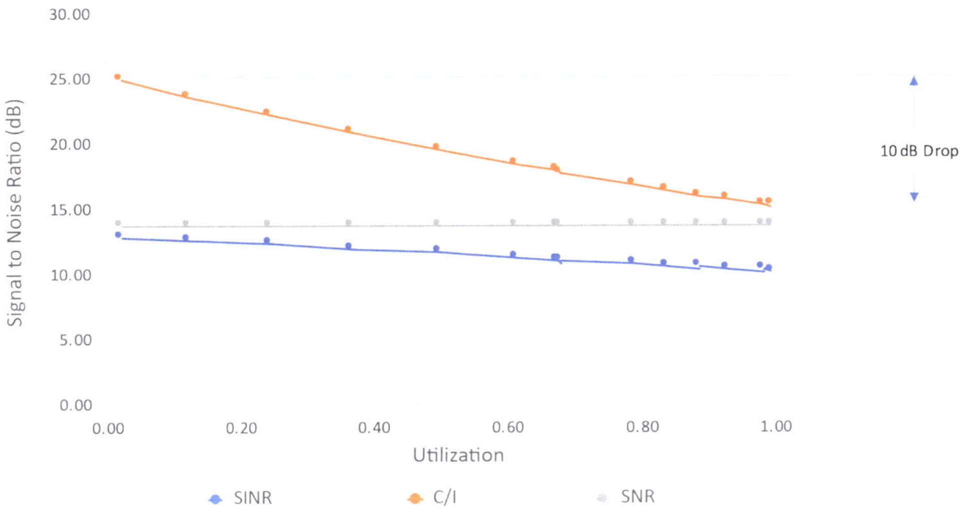

Figure 22: *Network signal-to-noise ratios vs network Resource-Block utilization*

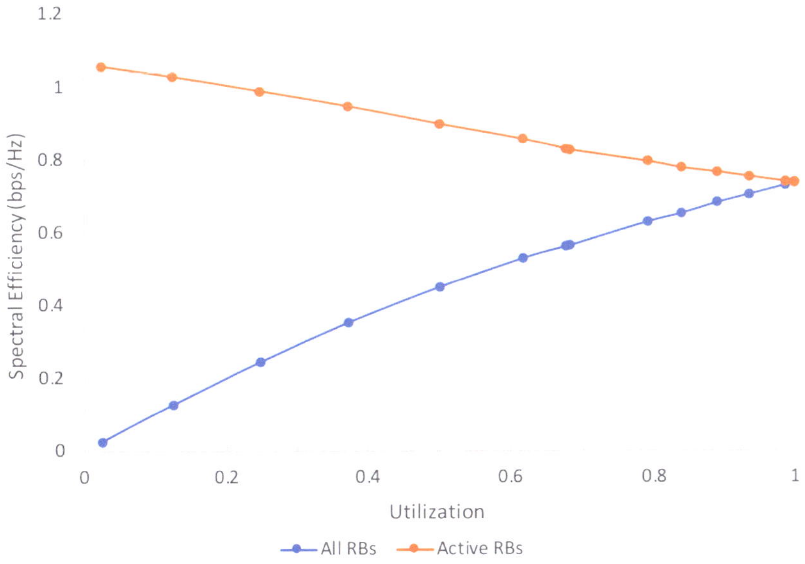

Figure 23: *Network spectral efficiency vs network Resource-Block utilization.*

40 Recall that spectral efficiency is defined as traffic divided by spectrum bandwidth. Thus, as traffic increases, spectral efficiency increases.

The above graphs display network KPIs. Similar graphs can be produced to display KPIs for individual cell sites. However, I would be very careful in interpreting the graphs of cell-site KPIs. For example, below is the analogous graph of the Network UPT versus Network Traffic (Figure 20) but for individual cell sites (Figure 24).

Figure 24: *Cell-site user-perceived throughput vs cell site traffic.*

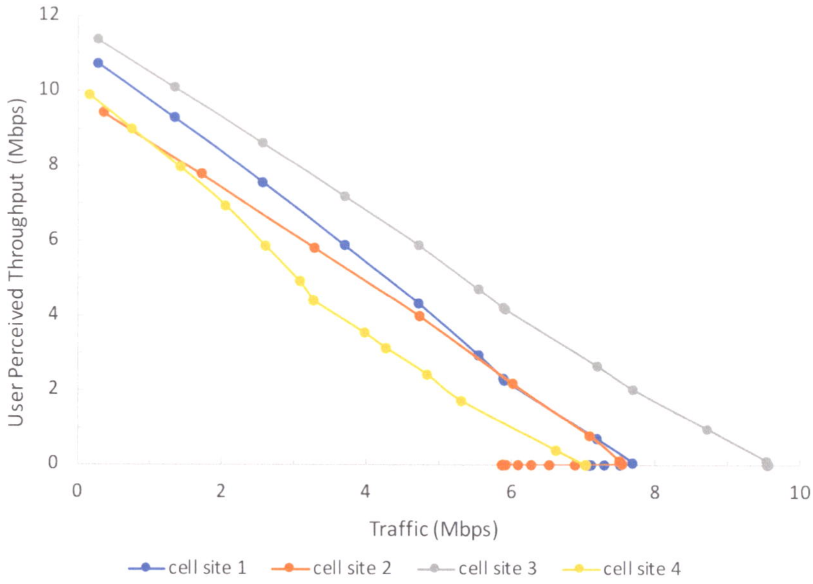

Looking at Figure 24, it may seem cell site 4 has some serious performance issues. At any given traffic level, the UPT of cell site 4 is much lower than any other cell site. Perhaps a cell-site technician may have seen this problem when analyzing cell-site KPIs of a real network. The tech might employ significant resources to find out why cell site 4 is performing so poorly, try to collect data by scheduling a series of expensive drive tests, or look to spend time redesigning cell site 4. Maybe the cell-site technician will receive praise for spotting the underperforming cell site. A business-case analysis may show that improving the performance of cell site 4 will save the company a bunch of money, hence providing justification for an allocation in next year's budget to fix the problem.

This is one of the problems with using data to prove a point. The person who has access to data can present it in a manner that best serves an agenda. Hence, the analysis of data is not always objective. In the above scenario, I doubt the cell-site technician would have anything to gain in allocating resources to improving the performance of cell site 4. In this situation, the attempt is mainly misguided. However, the above graph and analysis does seem convincing.

To refute the claim that cell site 4 is underperforming, inspect the following graph.

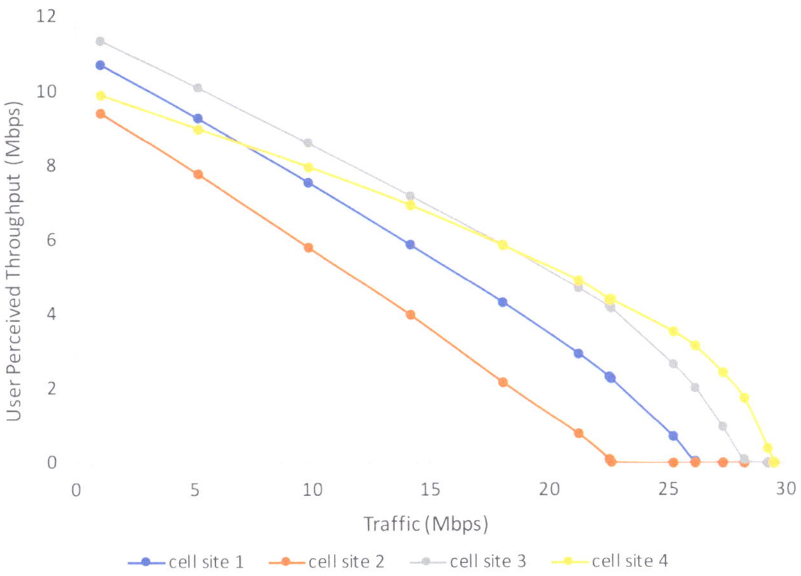

Figure 25: *Cell-site user-perceived throughput vs network traffic.*

Instead of plotting the cell-site UPT against cell-site traffic, Figure 25 plots the measure against network traffic. Now the UPT of cell site 4 doesn't look so bad. Under heavy network loading, subscribers in cell site 4 receive the best QoE of all of the cell sites. It is cell site 2 that is the problem. The reason is quite simple. Cell site 2 has the most subscribers, hence it supports the most traffic. It is the cell site that gets congested first. This is the cell site that the technician should have been focusing on.

Before leaving this example, I would like to refer back to Figure 24. We know that cell sites 1 and 3 have the same number of subscriber locations and thus transmit the same amount of data. However, cell site 3 has better UPT than cell site 1 at any given traffic level. The key factor that determines how quickly a sector experiences congestion is the number of subscribers served by the sector. When two sectors serve the same number of subscribers, the locations of the subscribers will play a role in determining which sector will congest sooner.

Another important observation is that the maximum capacities of each of the cell sites are different. Recall that the network planner using the heuristic method assumes that all sectors have the same maximum capacity. Inspecting Table 21 and Figure 24, not only are the capacities of each of the cell sites different, in the case of cell site 2, its maximum capacity (7.5 Mbps) occurs when network Resource Block utilization is at only 68%. At 100% network Resource Block utilization, the amount of traffic supported by cell site 2 is only 5.87 Mbps. At 68% network utilization, cell site 2 utilization is at 100%, which implies cell site 2 is transmitting data continuously. As network utilization increases beyond 68%, cell site 2 is still transmitting data continuously, but subscribers in cell site 2 experience greater interference from other cell sites operating at higher Resource-Block utilizations, thus impacting data-link speeds for cell site 2 subscribers. The figure below shows the amount of traffic carried by each cell site based on the level of network utilization.

Figure 26: *Cell-site traffic versus network utilization.*

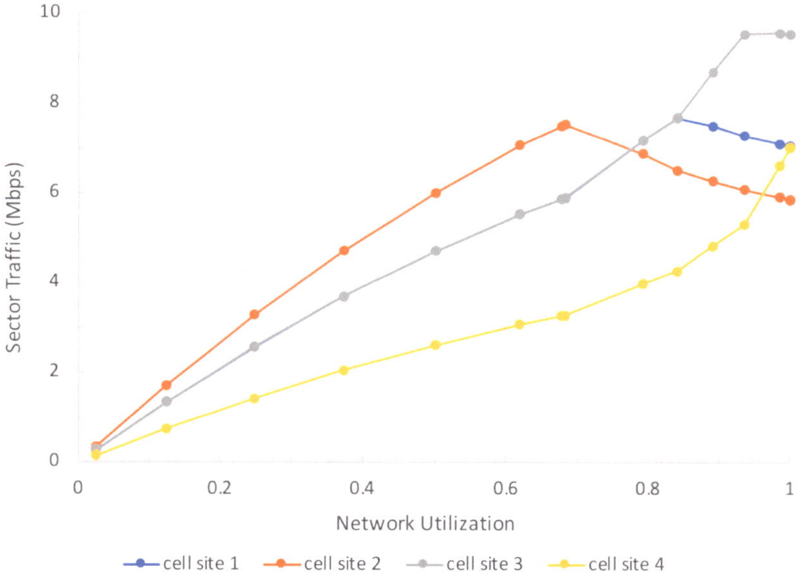

Note that the traffic on cell site 2 reaches its maximum at 68% network utilization. Also, below 68% utilization the ratio of traffic between cell site 2 and cell site 3 is the same as the ratio of subscriber locations for the two cell sites (23:18). Cell site 1 and 3 carry the same amount of traffic below 83.5% network utilization, which explains why you don't see the graph for cell site 1 until it becomes congested. Above 83.5% network utilization, cell site 1 is at 100% Resource-Block utilization and sector traffic can only decrease as co-channel interference increases with higher network utilization. At 94.5% network utilization, cell site 3 becomes congested. Cell site 4 is the last to congest at 100% network utilization. Although cell site 1 has 1.8 times as many subscriber locations as cell site 4, they both carry the same amount of traffic at 100% network utilization and more traffic than cell site 2, which has more subscriber locations than either cell site 1 or 4. The amount of traffic carried by a sector is not a function of the number of subscriber locations in the sector when all sectors are at 100% utilization.

3.4.4.1 Selecting the Number of Subscriber Locations

If one is very observant, the graph for the fifth percentile cell-edge UPT in Figure 20 and the graph for cell site 2 UPT in Figure 25 are identical. Essentially, the former is the user-perceived throughput of cell site 2. From Table 21, one can see that cell site 2 has the most subscriber locations and hence the most traffic.[41] Using Equation (11), the greater the amount of traffic, the slower the UPT. Although it is an impossible task, the cell-site engineer will try to have each cell-site sector transmit the same amount of data. In the simple network example, the fact that only 69 subscriber locations were randomly placed means that it was very likely that one sector would receive meaningfully more subscribers than the others (in this case, cell site 2). Hence, limiting the number of subscribers per sector partially mimics the traffic imbalance observed in an actual network. The greater the traffic imbalance between cell-site sectors, the lower the cell-edge UPT. Had we placed significantly more than 69 random subscriber locations, you would find that the number of subscribers in each sector will become more similar and less varied, which is not typically observed in an actual network. The figure below shows the UPT versus traffic graph with 40 times more subscriber locations used in the simulation.

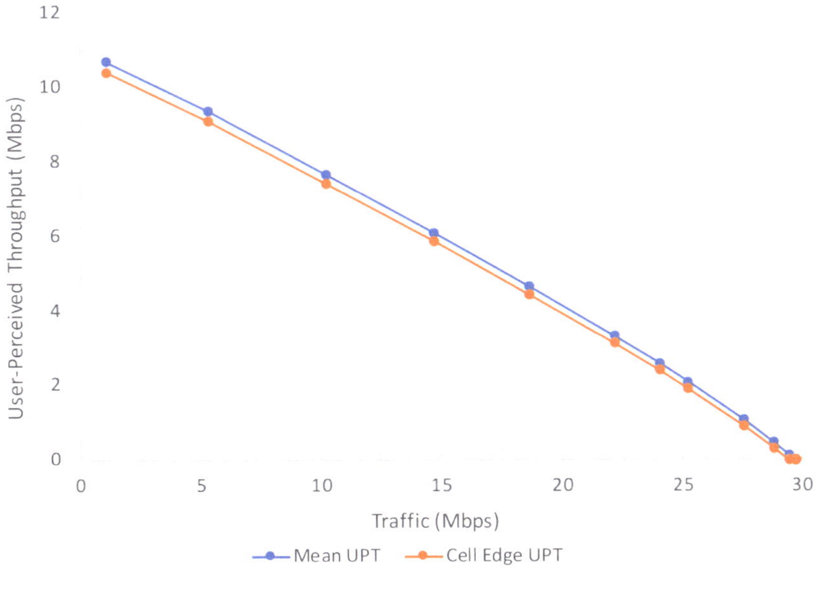

Figure 27: *User-perceived throughput vs traffic for approximately 2,900 subscriber locations.*

41 Recall each subscriber location downloads the identical amount of traffic unless a sector is in congestion.

The numbers of subscriber locations in each sector are nearly identical: 715, 740, 701 and 724. As a result, each cell has approximately the same amount of traffic and the cell-edge UPT nearly matches the mean UPT. This is in contrast with Figure 20, which represents what is typically observed in actual wireless networks.

Compare the graph below with Figure 24.

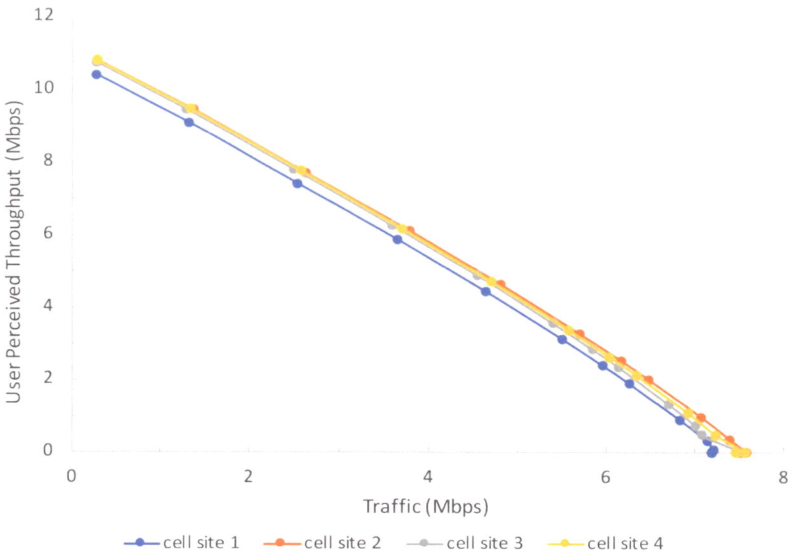

Figure 28: *Cell-site UPT vs cell-site traffic.*

This is the graph which might have been anticipated due to the symmetry of the network. (See Figure 16.) The graphs for each cell site are nearly identical, since they have nearly identical number of subscriber locations. The variations in the traffic and UPT graphs for each sector as observed in Figure 24 arise because the number of subscriber locations is only 69.

Later in this book, when we look at small cells, we will use other techniques to create this sector-traffic imbalance. For now, randomly placing subscribers such that the average number of subscribers per sector is around 20 will create enough variation in the number of subscribers per sector to mimic the traffic imbalance seen in actual networks.

Why is this important? Operators are always concerned with the UPT of the subscribers in the busiest sectors since, once those sectors congest, the network is considered to be at capacity. In the stochastic method, the busiest sector is the one with the most subscriber locations, since it is assumed that each subscriber location downloads the same amount of data if there are no sectors in congestion (i.e. no sector at 100% Resource Block utilization). If the number of subscriber locations are unevenly distributed amongst the sectors, then it is likely that there could be much unused capacity in the non-congested sectors when the network reaches capacity (i.e. the busiest sector has reached 100% utilization). As mentioned above, the busiest sector defines the cell-edge UPT, and it is desirable that it not be significantly lower than the mean UPT. It is unrealistic that the cell-edge UPT be as close to the mean UPT as shown in Figure 27. The cell-edge UPT in Figure 20 is a more realistic goal for network planners.

3.4.5 The Importance of η

In the first paragraph in the section 3.4.2 "Analyzing Subscriber KPIs" I state that all simulations are performed with the adjusted thermal noise floor, η, set to 10. We will now discuss the implication of setting η to various values. Remember, the definition of η, stated in (48), is:

$$\eta = 10^{(rnf - Transmit\ Power - Gain + Misc\ Loss + 45.5 + (35.46 - 1.1h_m)\log f_c + 0.7h_m + C + 10\gamma \log \tau)/10}$$

η does contain the receiver noise floor (*rnf* in dBm), and the equation $10^{rnf/10}$ is the standard definition of thermal noise used in Equations (36) and (37), which are the more recognisable definitions of SINR. However, Equation (48) has an assortment of other terms.

- h_m is the height of the mobile antenna in meters
- f_c is the frequency of the carrier in MHz
- C is a correction factor
  - C = 0dB for suburban and rural
  - C = 3dB for urban environments
- Transmit power of the radio specified in dBm
- Gain provided by antennas in dB
- Miscellaneous losses associated with connectors, cables, building material, body loss, etc.
- γ is the radio frequency path-loss exponent expressed in dB/decade/10
- τ is the distance scaling factor first discussed in Equation (45).

When η is set to 10, then:

$$10 * \log \eta = rnf - Transmit\ Power - Gain + Misc\ Loss + 45.5 + (35.46 - 1.1h_m)\log f_c + 0.7h_m + C + 10\gamma \log \tau = 10\ dB. \quad (54)$$

The table below shows various combinations where η = 10. Note that in Figure 16, the inter cell-site distance was a nominal 1,750 meters. To get η = 10, the scale factor τ has to be adjusted, thus affecting the inter cell-site distance for each of the combinations.

Combinations	1	2	3	4
rnf (dBm)	-97.00	-97.00	-97.00	-97.00
Transmit Power (dBm)	36.99	36.99	40.00	40.00
CPE Antenna Gain (dB)	0	0	0	0
Base Antenna Gain (dB)	6	6	6	0
Total Gain (dB)	6	6	6	0
Miscellaneous Losses (dB)	20.00	20.00	20.00	20.00
Frequency (MHz)	2000	850	2000	2000
CPE Antenna Height (m)	1.5	1.5	1.5	1.5
C	0	0	0	0
Path-Loss Exponent, 10γ	40	40	40	40
Scale Factor, τ	0.20	0.41	0.23	0.17
Noise Factor, η (dBm)	10.00	10.00	10.00	10.00
Noise Factor, η	10.00	10.00	10.00	10.00
Inter cell-site distance (m)	345.81	712.72	411.24	291.15

Table 22: *Various scenarios where $\eta = 10$. The shaded boxes indicate changes from combination 1.*

All the above combinations will produce the same simulation results shown thus far in the chapter. This is the labour savings I've been touting; perform the simulation once, and the results are applicable for an infinite number of scenarios.

Note that reducing the operating frequency from 2,000 MHz in combination 1 to 850 MHz in combination 2 increases the inter cell-site distance by a factor of two. This means with the same number of cell sites, you can cover four times the area using 850 MHz versus 2,000 MHz.

Note that η is common for all subscriber locations. What is unique for each subscriber location is the term $10\gamma\log(d/1000) + F$ found in Equation (49) and other similar equations. In many situations, η is relatively small compared to other terms in Equation (49) and can be ignored. This scenario is characteristic of dense urban environments where the SINR for a subscriber location is essentially equal to *C/I*. In situations where η is not insignificant, it affects both capacity and coverage. We will be adjusting η and observe the effects it has on coverage/capacity in the following sections.

3.4.5.1 Scenario 2: Minimum-Required SNR = 2, η = 5

I changed η from 10 to 5 but, to keep the coverage the same, increased the minimum-required SNR from 1 to 2. Recall that to be in coverage, the desired signal must be larger than η * minimum-required SNR. In Scenario 1, η = 10 and minimum-required SNR = 1. In Scenario 2, η = 5 and minimum-required SNR = 2 (refer back to the section "Coverage Analysis" if there is any confusion). Just like the original scenario, only 69 of 80 subscriber locations meet the coverage criteria in Scenario 2. In fact, the exact same subscriber locations meet the minimum-required signal strength. So, not only are the geometry of the cell sites unchanged, the geometry of the subscriber locations is also unaffected. This minimizes the changes between Scenario 1 and Scenario 2. When we change a parameter (in this case, the minimum-required SNR) it is easier to observe its effects on network KPIs when everything else remains unchanged. If, for example, different subscriber locations were used in Scenarios 1 and 2, then we wouldn't know what changes to network KPIs were due to changes in the minimum-required SNR or to different subscriber locations used in each scenario. Alternatively, we could have selected subscriber locations deliberately and not have them disqualified by minimum-required signal strength. This means in our simple example, the same 80 subscriber locations would be used in all scenarios regardless of signal strength or SNR requirements.

There are many ways to change η from 10 to 5. The easiest is to double the base-station power. Another is to increase the antenna gain by 3dB at either the customer-premises equipment (CPE) or base station. Reducing the inter cell-site distance by 16% would also work. Reasons for increasing the downlink minimum-required SNR from 1 to 2 might be due to a change in the air-interface standard, or due to a limitation in the uplink. (See section 3.3.1.)

The easiest method to observe the effects of changing η is to plot graphs for η = 10 and η = 5 on the same chart.

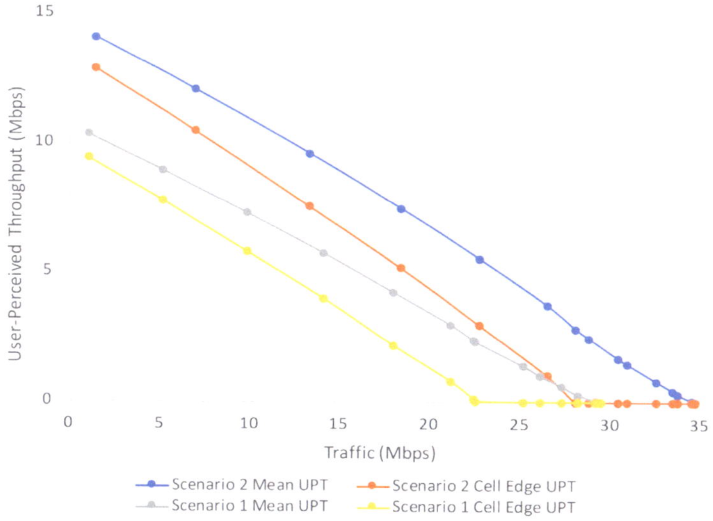

Figure 29: *User-perceived throughput vs traffic for Scenarios 1 and 2. Scenario 1: η = 10. Scenario 2: η = 5.*

There is an obvious improvement in capacity and UPT when decreasing η by 50%.

The various signal-to-noise ratios are shown in the graph below.

Figure 30: *Network signal-to-noise ratios vs network Resource Block utilization for Scenario 2.*

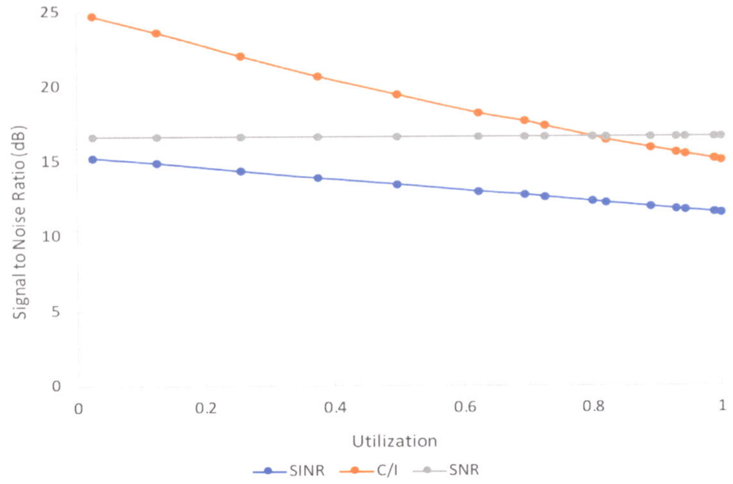

Note that the SNR curve is 3 dB higher ($10*\log_{10}2$) in Figure 30 versus the SNR curve in Figure 22. Also note that the *C/I* curves are identical in Scenario 1 and 2. *C/I* does not depend on η.

3.4.5.2 Scenario 3: Minimum-Required SNR = 4, η = 2.5

Continuing in the same vein, we will reduce η by another 50% and double the minimum-required SNR. Once again, the same subscriber locations meet the coverage criteria.

The graph below compares the UPT vs traffic for Scenarios 1 and 3.

Figure 31: *User-perceived throughput vs traffic for Scenarios 1 and 3. Scenario 1: η = 10. Scenario 3: η = 2.5.*

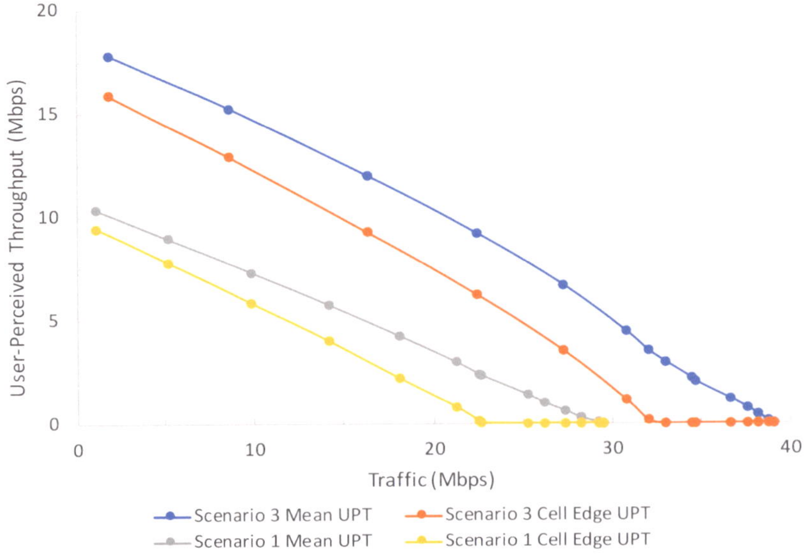

A pattern is emerging. The lower the noise η, the better the network performance. The impact of co-channel interference is becoming more evident, and that is a good thing.

The various signal-to-noise ratios are shown in the graph below.

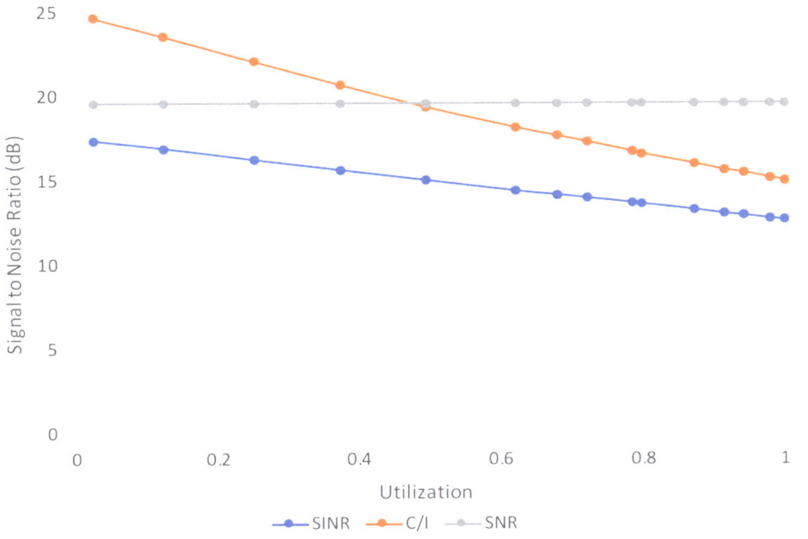

Figure 32: *Network signal-to-noise ratios vs network Resource Block utilization for Scenario 3.*

Note that the SNR curve is 6 dB higher ($10*\log_{10}4$) in Figure 32 versus the SNR curve in Figure 22, and 3 dB higher versus the SNR curve in Figure 30. Also, note that the *C/I* curves are identical in Scenarios 1, 2 and 3.

3.4.5.3 Scenarios 4 & 5

Rather than producing the same graphs for Scenarios 4 and 5, I combined all the scenarios on one graph. First, I'm going to summarize the parameters of the scenarios in the following table:

Scenario	η	Minimum-Required SNR	Minimum-Required Signal Strength	Mean SNR (dB)
1	10	1	10	13.66
2	5	2	10	16.67
3	2.5	4	10	19.68
4	1	10	10	23.66
5	0.1	100	10	33.66

Table 23: *Summary of parameters for Scenarios 1 to 5.*

The first graph shows the mean UPT vs traffic for each of the scenarios. Reduce the value of η, improve the performance curve.

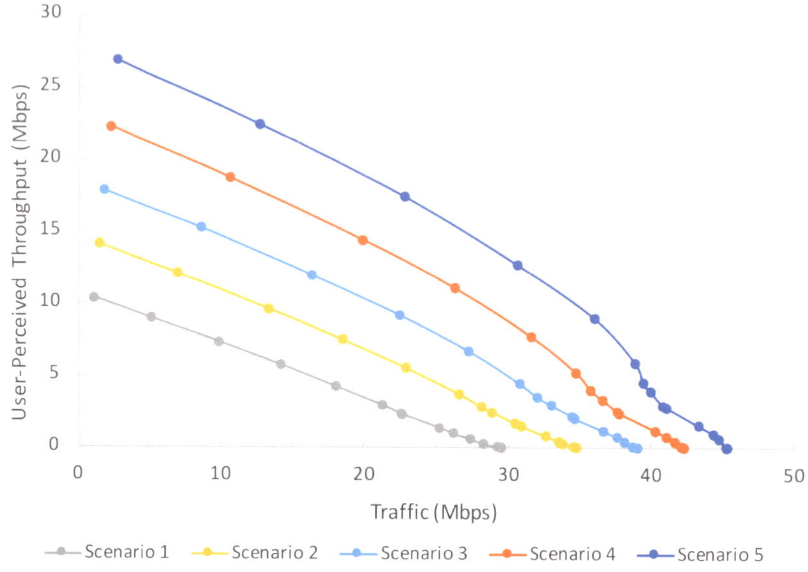

Figure 33: *Mean user-perceived throughput vs traffic.*

The graph below shows the SINR vs utilization for each of the scenarios. The *C/I* is also displayed, which is common for all the scenarios as a reference. Referring to Table 23, the last column displays the mean SNR for each of the scenarios. The higher the mean SNR, the closer the SINR curve follows the *C/I* curve. In Scenario 5, when the mean SNR is

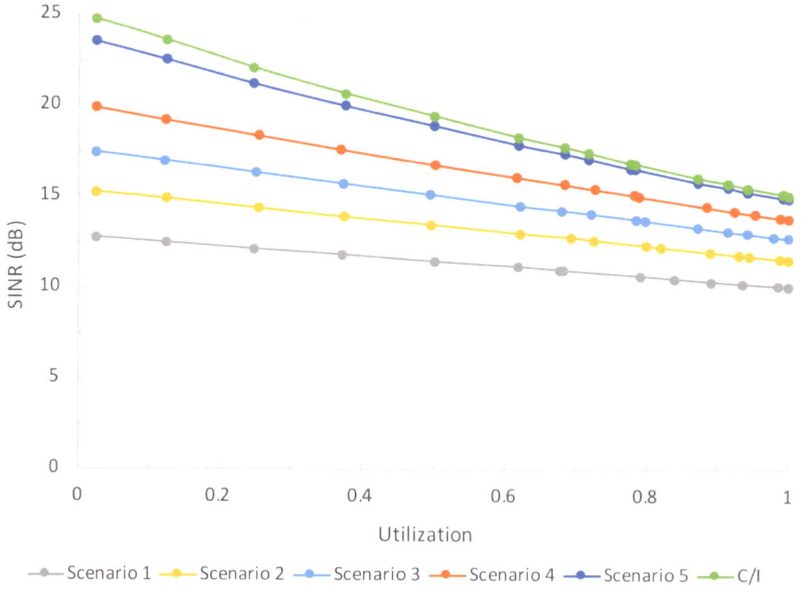

Figure 34: *SINR vs utilization.*

much higher than the *C/I*, the SINR curve almost matches the *C/I* curve. Thus, one can say Scenario 5 is strictly co-channel interference limited, even under low traffic loads. On the other hand, Scenario 1 is much more affected by thermal noise, although I wouldn't say it is strictly noise limited since co-channel interference does affect network performance under higher traffic levels.

3.5 Constant-Interference Scenario

The impacts of co-channel interference have been noted numerous times, not the least of which is its role in complicating the analysis of wireless networks vis-a-vis wireline networks. Because of co-channel interference, we see in Figure 21 that the mean UPT versus traffic graph is steeper for the wireless network than for the wireline network. We see in Figure 22 that SINR decreases as network utilization increases, resulting in a lower average spectral efficiency for active Resource Blocks, as displayed in Figure 23. The impact of co-channel interference is clearly seen in Figure 26, where sector traffic decreases with increasing network utilization for cell sites that are at 100% utilization (cell sites 1 to 3).

It was mentioned in section 2.8 that by allocating each cell site exclusive spectrum, the impact of co-channel interference will be eliminated. There is, however, another method to decrease the impact of increasing co-channel interference from growing network utilization without giving each cell site its own spectrum allocation. If all cell sites constantly transmit at full power regardless of whether there is data to be transmitted, then the amount of co-channel interference will not change as network utilization increases, even if all cell sites occupy the same spectrum band. Referring to Equation (49), the term ABS_i is set to 0 dB regardless whether or not there is data to be transmitted.

Although this strategy appears counterproductive, at 100% utilization the capacity of this network and the network analyzed in the previous section are the same, since there is nothing to tell them apart. The networks differ only when there are Almost Blank Subframes (ABS). There must be some penalty for radios to be transmitting noise in empty Resource Blocks, but it won't be ascertained if the only KPI of interest is the amount of network traffic supported when all LTE subframes are carrying data. This also implies that the quality of the experience of the subscriber is unimportant, since UPT—its most important KPI for subscribers—is always 0 Mbps at 100% Resource Block utilization.[42]

Revisiting Scenario 1 ($\eta = 10$, minimum-required SNR = 1 or 0 dB), we will compare KPIs of the network where radios transmit only the cell-

42 As an aside, if one allocates each cell site its own frequency band, it wouldn't matter if the radio transmits continuously. Point-to-point microwave radio links operate in this manner.

specific reference symbols (CRS) in the absence of data to the network where radios transmit energy in every Resource Element regardless if it carries data. If you are wondering what modifications are necessary to simulate these two scenarios, you will be happy to know that by modifying one variable—ABSoffset in the sample MATLAB code in section 3.4.1.3—we can set the power of the empty Resource Block to any value. For the scenario where the CRS is transmitted in empty Resource Blocks, ABSoffset is set to 10, indicating the power to transmit an empty Resource Block is 10 dB lower than a Resource Block carrying data. For the scenario where the radio is constantly transmitting, ABSoffset is set to 0, indicating the power to transmit an empty Resource Block is the same as one carrying data.

The figure below shows that constant transmission does have some impact on average UPT but no impact on the maximum network capacity.

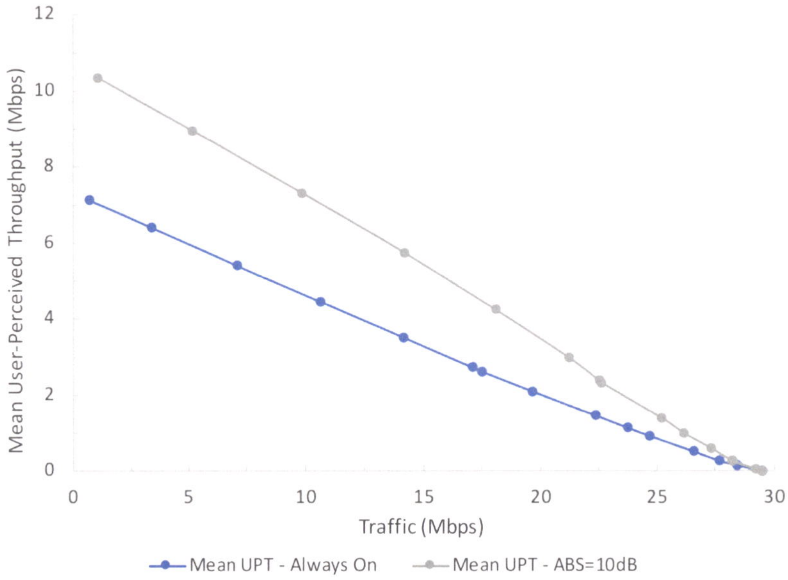

Figure 35: *Mean UPT versus traffic for regular LTE (ABS=10dB) and the network where radio power is constant.*

The reason for this less-than-dramatic change in the mean UPT versus traffic graph is that KPIs in Scenario 1 are affected more by thermal noise than by co-channel interference. If maximum network capacity is the only concern for network operators, it is difficult to assign any value in reducing the impact of co-channel interference via the suppression of radio transmission in the absence of data. However, MNOs are also concerned about subscriber QoE and rarely do they operate networks anywhere near 100% Resource Block utilization, due to the impact on UPT. Although the above figure does provide some support for reducing co-channel interference by suppressing unnecessary radio transmissions, a more compelling argument can be made.

Instead of focussing on average UPT, we examine cell-edge UPT, as suppressing unnecessary radio transmissions increases network capacity. (See figure below.)

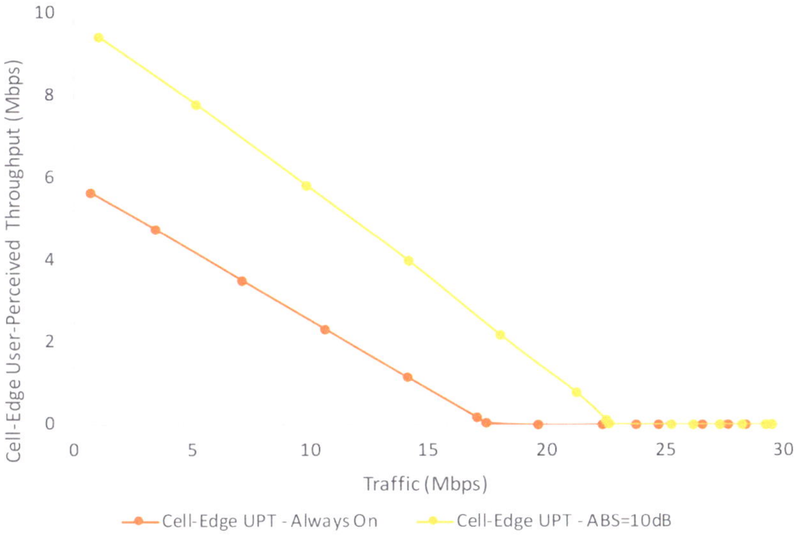

Figure 36: *Cell-Edge UPT versus traffic for regular LTE (ABS=10dB) and the network where radio power is constant.*

If network capacity is defined as the minimum amount of network traffic required to drive cell-edge UPT to 0 Mbps, then suppressing unnecessary radio transmissions will increase network capacity by 30% (22.6 Mbps versus 17.5 Mbps).

To gain further insight into the negative effects of continuous transmission, we will examine the sector traffic versus network utilization graphs shown below.

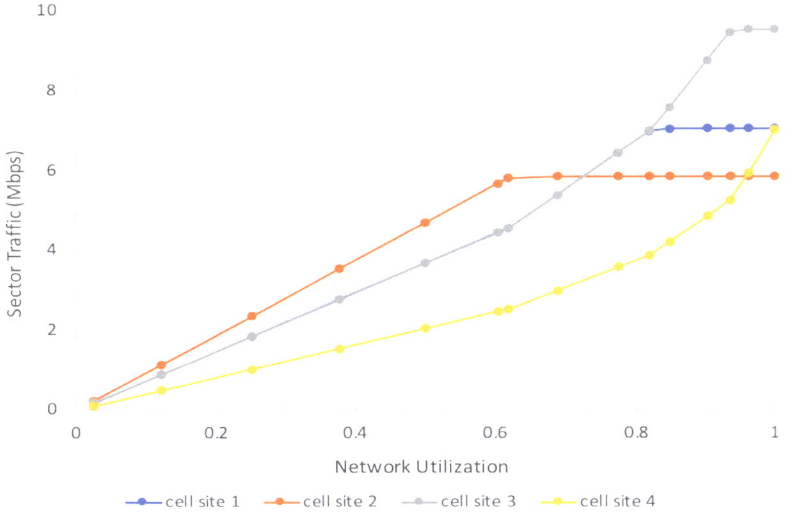

Figure 37: *Sector traffic versus network utilization for continuous radio transmission.*

If you examine cell-site traffic at 100% utilization, you will find the same values as in Figure 26 at 100% utilization. The two figures have obvious differences:

- In Figure 26, the curves do not consist of a series of straight lines. In the region where no cell sites are in congestion (<67% network utilization), all lines bend downwards. As utilization increases, co-channel interference increases, decreasing the average spectral efficiency of the Resource Blocks. In Figure 37, all curves are piecewise linear. Since co-channel interference is constant, the average spectral efficiency of Resource Blocks does not change. Thus, traffic increases linearly with utilization until there are no idle Resource Blocks, at which point sector traffic remains constant. Figure 37 looks similar to Figure 12, which shows link traffic versus network utilization for a fixed wireline network, except that the amount of traffic supported by cell sites is not identical, whereas each fixed link identically supports 10 Mbps.
- In Figure 26, cell site 2 traffic peaks at a higher value (30% greater) than in Figure 37, before declining to the same value at 100% utilization. This shows the benefit of suppressing unnecessary radio transmissions, as lower levels of co-channel interference result in higher data-link speeds and sector traffic. In a certain sense, suppressing unnecessary radio transmissions allow cell site 2 to steal capacity from less busy cell sites.

Analyzing Scenario 5 (η = 0.1 and minimum-required SNR = 100), suppressing unnecessary radio transmissions increases network capacity by 60%. Co-channel interference plays a bigger role in Scenario 5, which explains the bigger benefit for only transmitting the CRS in idle Resource Blocks.

As mentioned several times, the constant co-channel interference scenario is amenable to analytical or deterministic techniques, since SINR and data-link speed s_i are constant for each subscriber location, and Equations (6) and (7) can be used to calculate sector capacity and sector traffic. All the graphs in this section were established using the stochastic algorithm described in the chapter, but they could have been equally developed deterministically. This does not rigorously prove the stochastic approach is correct, but it is another indication of its credibility.

3.6 The Effects of Cell-Specific Reference Symbols

In the last section, we see the benefits of transmitting only the CRS when Resource Blocks have no data. Perhaps there are additional benefits by suppressing the CRS in idle Resource Blocks. If the CRS is eliminated, it is still possible to send reference signals to help with channel estimation and demodulation. These non-CRS reference signals are transmitted in occupied Resource Blocks for the specific benefit of the user equipment receiving the data. Mirroring the discussion in the previous section, at 100% utilization it can be argued that both networks (one with CRS and one without) will have identical KPIs, as differences will only be observed when there are a significant number of unused Resource Blocks. One may speculate whether 3GPP fully understood the implications of transmitting the CRS in idle Resource Blocks since the focus was on maximizing capacity where there are no idle Resource Blocks. The fact that the CRS is no longer present in 5G NR suggests that engineers had second thoughts about its benefits.

In this section, we will analyze the implications of not sending the CRS during idle Resource Blocks. For those interested in the sample MATLAB code in section 3.4.1.3, ABSoffset should be set to a high value: 100 dB for example. This implies that the power of an empty Resource Block is 100 dB lower than the power of a Resource Block with data.

Of the five scenarios previously analyzed, I choose Scenario 5 ($\eta = 0.1$ and minimum-required SNR = 100) since it produces the most interesting results. (See Figure 38.)

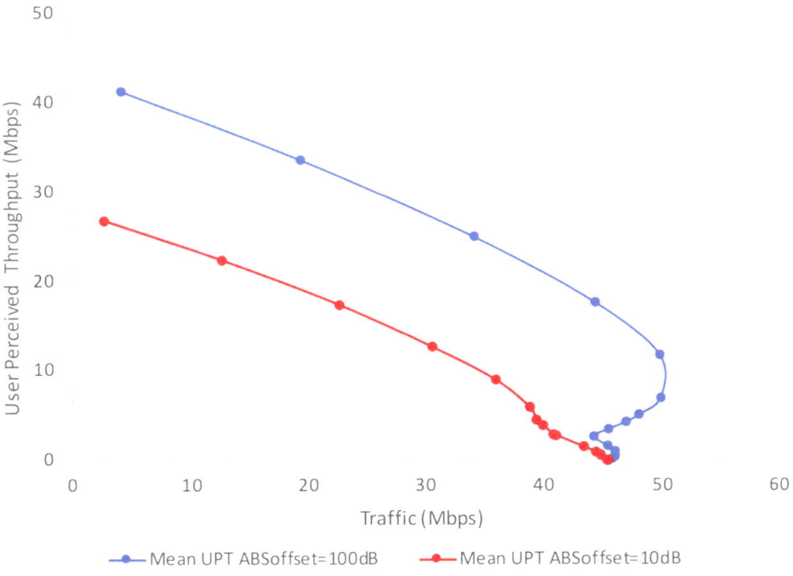

Figure 38: *Scenario 5 average UPT vs traffic with and without the CRS.*[43]

Analyzing the Mean UPT vs Traffic graph when the CRS is present (the bottom curve of Figure 38) we see the familiar trade-off between

43 CRS is present when ABSoffset=10dB and absent when ABSoffset=100dB.

network traffic and mean UPT. The trade-off seems reasonable and intuitive: the greater the traffic, the slower the service. There are four major points regarding the Mean UPT vs Traffic graph when the CRS is absent (the top curve of Figure 38).

- At 100% utilization, both curves of Figure 38 have the same amount of traffic and obviously the same UPT (0 Mbps). This illustrates that these curves differ when there are idle Resource Blocks.
- The mean UPT is much higher when the CRS is absent. Even though the power in idle Resource Blocks with CRS are 10 dB lower than active Resource Blocks, the co-channel interference generated by the CRS still has a significant impact on KPIs.
- There are certain parts of the graph without the CRS that are counter intuitive. Parts of the curve suggest that by increasing traffic we can have higher mean UPT. Let's assume that all resources are consumed (i.e. every Resource Block is carrying data). How is it possible to add more data and at the same time get a better user experience? What is really happening is that when utilization decreases, the average spectral efficiency of the occupied Resource Blocks increases faster. Also, UPT increases as the proportion of idle Resource Blocks increases.
- Without the CRS there are parts of the graph where there are two distinct UPTs for a given network traffic level.

The operator deploying a 5G network using omni-directional antennas in an environment where co-channel interference is dominant must be very careful not to let network utilization exceed 38%, else it runs the risk of UPT dropping quickly towards 0 Mbps. Note that, at 38% utilization, the amount of traffic on the 5G network is the same as the 4G network at 100% utilization. Thus, the biggest benefit of eliminating CRS is not capacity gain but the improvement in UPT.

Most papers assume that maximum traffic capacity occurs when resources are 100% utilized. The above figure demonstrates the hazards of making assumptions without providing proof. Figure 38 shows that when co-channel interference dominates, maximum traffic can occur at levels of utilization far below 100%. Also, it would be silly for an MNO to operate a network at 100% network utilization when it could carry the same amount of traffic at 38% utilization and have a much higher mean and cell-edge UPT. The key takeaway is that co-channel interference is not evil, but it must be managed. In the above scenario, co-channel interference limits network utilization to less than 38%. This point will be missed if the primary focus is on full-buffer simulations.

When thermal noise plays a larger role, as in Scenario 1, the benefits of not transmitting the CRS for idle Resource Blocks is muted, as shown below.

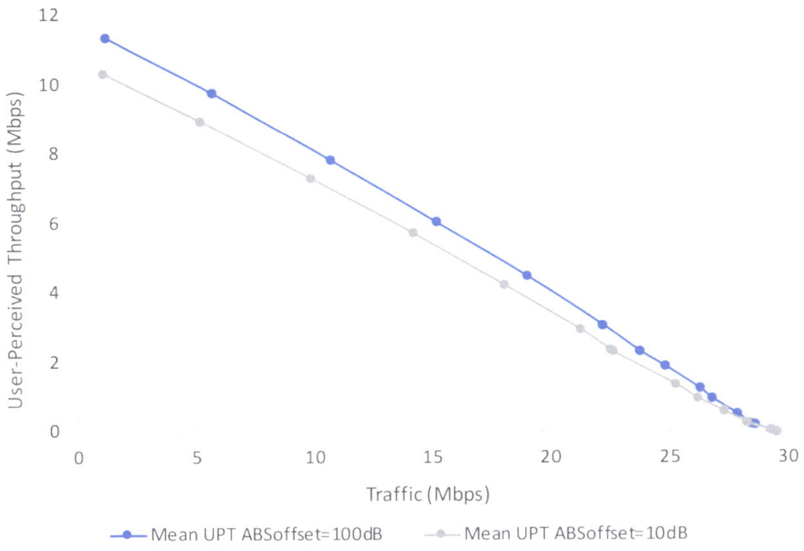

Figure 39: *Scenario 1 average UPT vs traffic with and without the CRS.*[44]

The inclusion of the CRS does not have a large impact on UPT, and the figure below demonstrates that capacity is not severely affected.

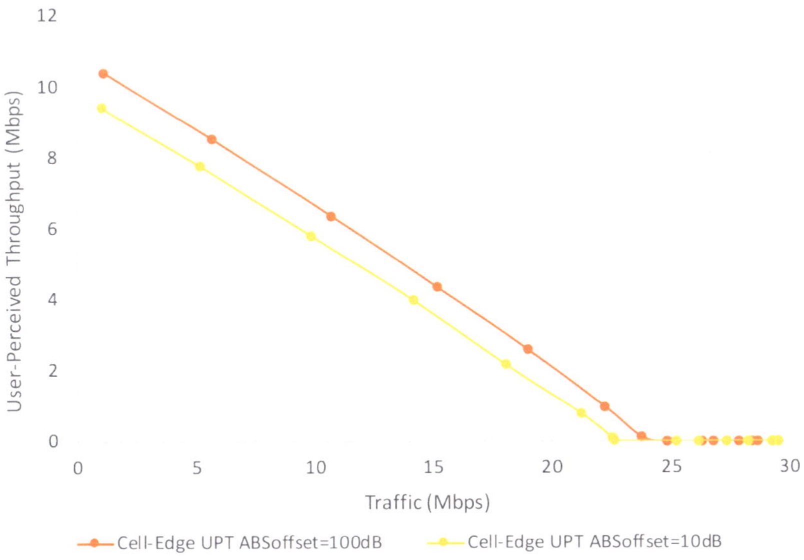

Figure 40: *Scenario 1 cell-edge UPT vs traffic with and without the CRS.* [45]

44 CRS is present when ABSoffset=10dB and absent when ABSoffset=100dB.
45 CRS is present when ABSoffset=10dB and absent when ABSoffset=100dB.

This concludes the discussion of the stochastic approach. Despite the simplicity of the toy network, had the stochastic approach not been used, several interesting insights may not have been discovered. Succeeding chapters will use the methods developed here to answer questions on the effectiveness of various wireless network designs and strategies. Many unexpected results were discovered in the process of extending the techniques presented in this chapter.

4 | Carrier Aggregation and Coordinated Multipoint

LTE-A (Long Term Evolution Advanced) introduced many features, but the main three are Carrier Aggregation (CA), Coordinated Multipoint (CoMP) and enhanced Inter-cell Interference Coordination (eICIC). I will analyze the first two features in this chapter while reserving the analysis of eICIC for later, when I discuss small-cell strategy. Without the stochastic approach developed in the previous chapter, it would have been impossible for me to analyze the impact of these features.

I had reservations whether I should attempt to analyze CoMP for this book. Trying to read the literature on CoMP, it is difficult to understand exactly how each of the various versions of CoMP work, never mind quantifying the benefits of each. Not to overwhelm the reader, I will only analyze one version. Rather than perform an exhausted analysis of CoMP, I will demonstrate how using the stochastic algorithm developed in the previous chapter makes analyzing a fairly complicated wireless feature such as CoMP relatively simple.

4.1 Carrier Aggregation

Carrier Aggregation is one of the first problems for which I used the model. CoMP and eICIC would each take a chapter to describe properly, but lowly Carrier Aggregation can be described in one sentence. CA bonds two or more separate spectrum bands to effectively act as one link. (See Figure 41 and Figure 42.)

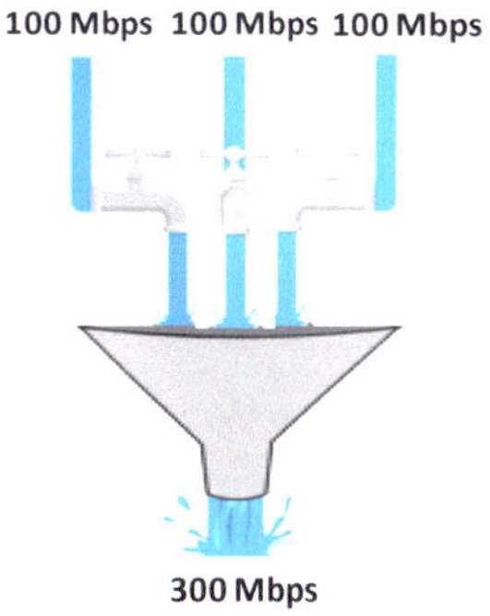

Figure 41: *Carrier Aggregation.*

Figure 42: *Downlink Carrier Aggregation.*

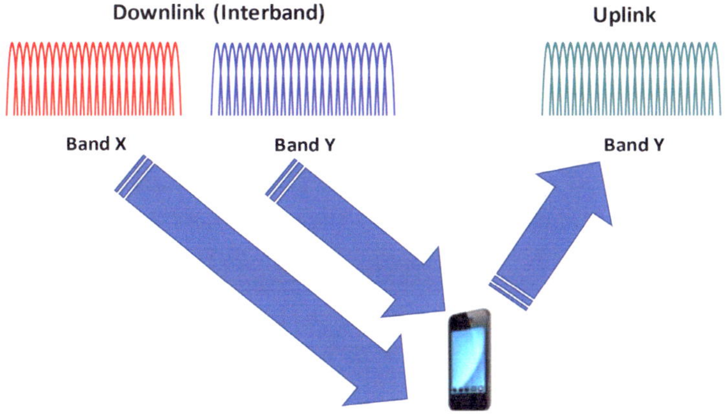

CA can be used for both the downlink and the uplink. The majority of the data traffic (>90%) is on the downlink so CA was developed for it first, as that would have the greatest impact.

An equivalent analysis problem is quantifying the advantage of having one spectrum band of 20 MHz versus two non-contiguous spectrum bands of 10 MHz each. If there is an advantage of a single contiguous spectrum block, then Carrier Aggregation could combine two non-contiguous spectrum blocks to appear as one.

Referring to Chapter 1, where two simple network dimensioning methodologies were introduced, both would attack the problem the same way. Both methods would assume that a certain amount of spectrum on a cell-site radio can support a certain amount of traffic. The heuristic method defines the maximum amount of traffic a 10 MHz radio can support while the alternative method defines the average amount of traffic a 10 MHz radio can support. Both methods would then double the amount of traffic for a 20 MHz radio. Instead of doubling the width of the spectrum band, one could add a second 10 MHz radio and the capacity of the wireless network would also double. It would be difficult to ascribe a value for Carrier Aggregation or contiguous spectrum because it appears both configurations had no advantage over non-contiguous spectrum without Carrier Aggregation. However, if you were to ask someone from Operations what they would rather have—one 20 MHz PCS spectrum block or two 10 MHz PCS spectrum blocks—most would choose the single block.

So, we are back at trying to determine the value of Carrier Aggregation and/or contiguous spectrum. It does not appear either provides any increases in capacity. Or does it?

To solve this problem of relative value, you could re-read Chapter 2 and the section titled "Capacity, Traffic and User-Perceived Throughput." However, you don't need to perform any simulations to solve this problem. In fact, if you stare at Figure 20 long enough, perhaps the answer will appear.

Determining the value of Carrier Aggregation and contiguous spectrum was a problem which was not initially difficult to solve, but the solution created a new set of problems. Take a look at Figure 4. It provides the LTE data-link speed given the SINR for 10 MHz of spectrum. What happens to the graph when the spectrum is doubled to 20 MHz? For those who guess that the data-link speeds will also double, give yourselves a pat on the back. If we double the data-link speeds in Figure 4, how would that affect Figure 20? If you guess that the UPT would double, then you would be half correct. The values on the horizontal axis (i.e. traffic) would also double. (See Figure 43).

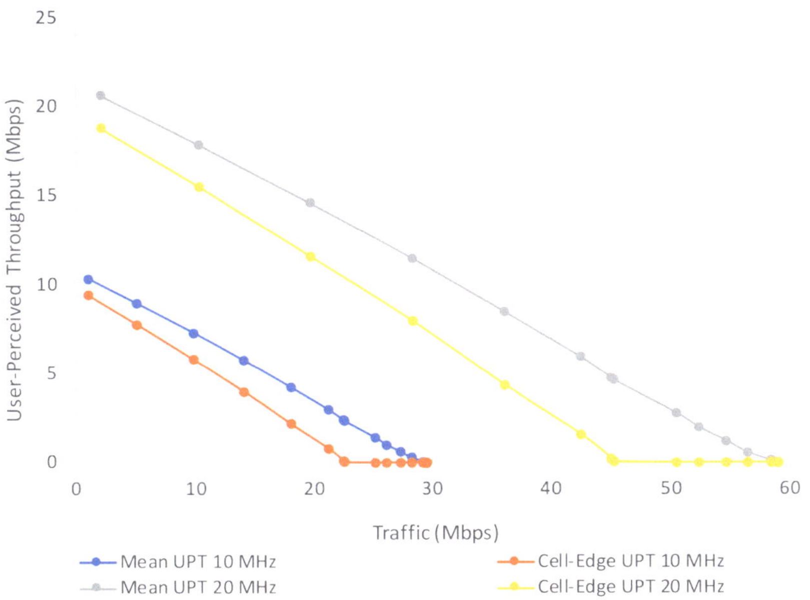

Figure 43: *User-perceived throughput vs traffic graph, 10 MHz vs 20 MHz.*

If the mobile network operator has two 10 MHz spectrum blocks, then the benefits of Carrier Aggregation can be seen in Figure 44.

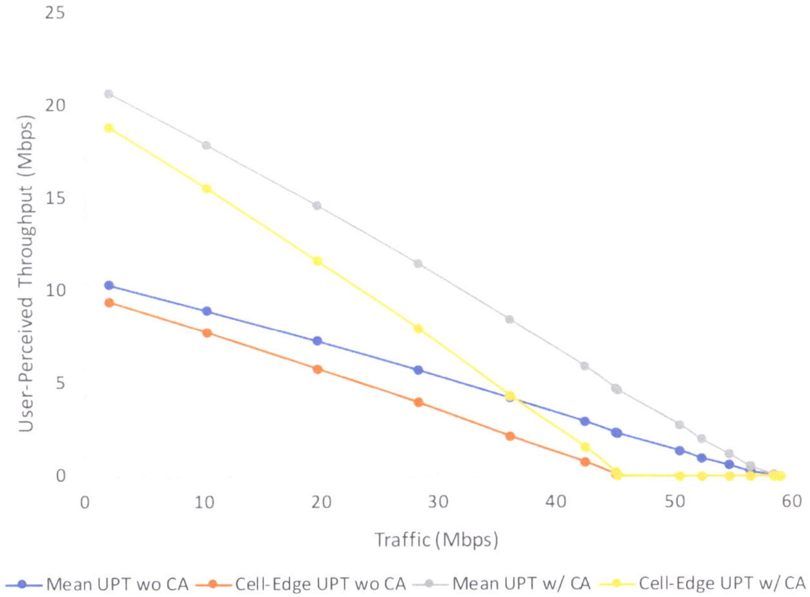

Figure 44: *User-perceived throughput with and without Carrier Aggregation. Two 10 MHz spectrum blocks assumed.*

For those wondering how the curves for mean and cell-edge UPT without CA were derived, use Figure 20 and simply double the numbers on the x-axis (since two 10 MHz are now being used), but the UPT numbers are for a single 10 MHz carrier (the y-axis numbers are unchanged). Since there are two 10 MHz carriers, the amount of traffic doubles, but a user can only use one of the carriers, hence the UPT is the same as for a single 10 MHz system.

If one focuses on the maximum amount of traffic that two 10 MHz blocks can support, then Carrier Aggregation doesn't provide any benefits, as both networks support the same maximum amount of traffic: 59 Mbps. However, at maximum traffic capacity, UPT is 0 Mbps. No mobile network operator would provision a network with such a poor Quality of Experience. The question is, how much traffic should be allowed? The answer may not rest solely within the authority of Operations. Marketing may also have a say as to how much traffic the network should carry, since it affects the public perception of the network. First, we have to decide whether to focus on the average or the cell-edge user-perceived throughput. Then we must choose the target (average or cell-edge) UPT.

SCENARIO 1:
So that we are not talking generalities, let's assume that Operations and Marketing decide that the 5th percentile cell-edge UPT should not fall below 2 Mbps. With no Carrier Aggregation, the amount of traffic in the simple network is 37 Mbps. If Carrier Aggregation is employed, network traffic is 41.5 Mbps, or 12% greater.

SCENARIO 2:
Assume that Operations and Marketing decide that the average UPT should not fall to less than 4 Mbps. With no Carrier Aggregation, the amount of traffic in the simple network is 37.4 Mbps. When Carrier Aggregation is employed, network traffic is 47.1 Mbps, or 26% greater.

Which scenario is more realistic? If you are an account representative for a mobile network equipment vendor, you would choose Scenario 2. You would market the feature as providing a 26% capacity gain without affecting the average UPT. Do you worry about the MNO buying 26% fewer base stations? Not really. MNOs have a set capital expenditure budget based on a percentage of the revenue. The profit margin on a software feature such as Carrier Aggregation is much higher than the margin on hardware.

If you are a cell-site planner working in Operations at an MNO, you worry about congested cell sites. With network traffic at 47.1 Mbps, there will be congestion with or without Carrier Aggregation. Hence, the conservative capacity gain is 12%, as outlined in Scenario 1. There is also improvement in the mean UPT. Referring to Scenario 1, without Carrier Aggregation and network traffic at 37 Mbps, the average UPT is 4.1 Mbps. With Carrier Aggregation and network traffic at 41.5 Mbps, the average UPT is 6.3 Mbps, or 54% greater. So not only is there a 12% increase in network capacity, there is a 54% increase in average UPT. The CAPEX benefit associated with the capacity gain is easy to model: a 12% capacity gain means that 12% fewer cell sites can be deployed while maintaining the cell-edge UPT target. The benefits of an increase in UPT is more difficult to determine.

Hence, it appears that the data support two different results, but both back the decision to implement Carrier Aggregation, lessening the impact of the ambiguous results.

4.1.1 The Role of Releasing Additional Spectrum Bands

The practice of using quantitative approaches for analysis can sometimes lead to unanticipated insights. User-perceived throughput is a relatively new KPI, which might explain why it does not get the same attention as traffic. This focus on traffic influences how people think about the role of additional spectrum.

The availability of spectrum has been an issue for the mobile wireless industry since its inception in the 1980s. In North America, government regulators granted spectrum licences to two companies in each area of the country. The wireline incumbent (the company that provided telephone service) received 20 MHz of spectrum, while a new entrant received the other 20 MHz. Not too long after launching service, MNOs started to experience traffic congestion as demand for mobile wireless service grew faster than anticipated. The solution was to grant each MNO an additional 5 MHz of spectrum. In the early days of cellular service, all traffic was voice. Each voice call required the same amount of spectrum bandwidth (30 kHz for AMPS). Doubling the amount of spectrum doubles the number of simultaneous voice calls that can be supported. Perhaps this is why people associate the availability of additional spectrum with supporting increasing traffic.

Every year, Cisco releases a report called the "Visual Networking Index (VNI) Global Mobile Data Traffic Forecast Update" that predicts future wireless data traffic growth. (See Figure 45 for an example.)

Regulators responsible for the allocation of spectrum (for example, FCC in the US, ISED in Canada) use this report and many others like it to guide them in establishing policies for releasing or re-purposing numerous spectrum bands. Although the principles developed in this

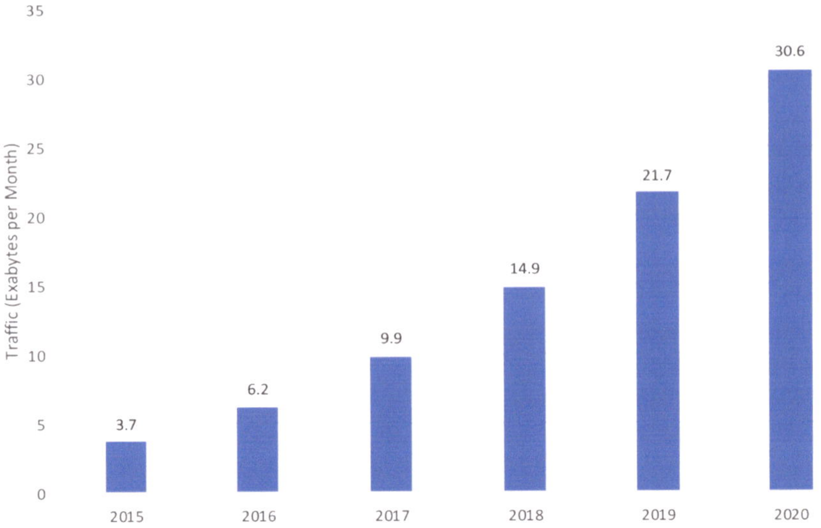

Figure 45: *Future data traffic.*[46]

book are appropriate for mobile, fixed, satellite and microwave backhaul applications, the focus of this discussion is mobile and consumer fixed-wireless access.

Regulators, cognizant of the growing demand for wireless data, see their role as accommodating growth by governing the amount of spectrum available to support such traffic. Wireless data forecasts assume that data growth will be exponential, since they all specify a compounded annual growth rate. (For example, Cisco predicted a compounded annual growth rate of 53% in Figure 45.) If the only mechanism to support future data traffic growth is the release of additional mobile spectrum bands, then the amount of spectrum allocated will also have to grow exponentially. Although there is a large amount of unused radio frequency bands, spectrum that is currently useful for mobile applications is limited and mostly allocated. The argument that additional spectrum is required to support traffic growth is not wrong, but incomplete.

The most obvious role that spectrum plays is its ability to support wieless data traffic, and the amount of spectrum plays a big role here. What doesn't get as much attention is the ability of spectrum to support data throughput as experienced by the subscriber. There is an implicit assumption when analyzing the capacity of the network that the user has sufficient data speeds for the application in use. Traditionally, the capacity or the amount of data traffic on the wireless network was the key metric, while the user-perceived throughput was given short shrift. However, we have just seen that UPT was key in understanding

46 2016 Cisco Visual Networking Index (VNI) Global Mobile Data Traffic Forecast Update.

the value of Carrier Aggregation.

The addition of spectrum bands should be less about supporting data-traffic growth and more about supporting applications requiring higher data speeds. Increasing the amount of spectrum will always support higher UPT if Carrier Aggregation is implemented, even if traffic congestion is not solved by its addition. The development of Carrier Aggregation changes the purpose of making available additional spectrum bands, from supporting increasing traffic to supporting higher user throughput. Of all the LTE-A features, Carrier Aggregation (used to increase data-link speeds) has provided operators greater benefits than other LTE-A features such as CoMP (Coordinated Multi-Point Operation) and eICIC (enhanced Inter-cell Interference Coordination), which were designed to increase capacity. To enable higher data-link speeds, Carrier Aggregation bonds spectrum across various frequency bands, thus making several non-contiguous spectrum bands appear as one larger contiguous band. Carrier Aggregation's impact on spectral efficiency (a measure of the technology's ability to carrying traffic) is secondary, although it now appears that Carrier Aggregation's ability to increase spectral efficiency is as good as other LTE-A features solely designed for that purpose.

We used UPT as a tool to derive the value for a particular LTE feature (Carrier Aggregation), but the analysis was incomplete as we didn't know how to value greater UPT. But the discussion on the impact of releasing additional spectrum bands suggests that the real purpose of Carrier Aggregation is to support applications requiring higher data speeds. Without the development of new applications requiring greater UPTs, there are other mechanisms besides the addition of new spectrum bands to support increasing data traffic. Other techniques, such as small cells or more advanced antenna technologies such as MIMO, can support growth in data traffic for existing applications.

4.2 Coordinated Multipoint

Several years after the launch of LTE, I was asked if there are new LTE features or techniques that can achieve greater spectral efficiency to support growing traffic demand. Knowing that co-channel interference is much more significant for wireless systems using N = 1 frequency reuse such as LTE than systems using higher frequency reuse such as WiMAX and Wi-Fi, I suspected that any feature designed to mitigate co-channel interference and increase data-link speeds would increase LTE network capacity. I suggested that we analyze a feature developed by 3GPP designed to reduce the impact of co-channel interference on LTE networks called Coordinated Multipoint (CoMP).

CoMP was not only touted as a method to increase spectral efficiency; it was also designed to increase the data-link speed of cell-edge users. Cell-edge users are defined by 3GPP as those with the lowest data-link speeds, which was assumed to be a result of greater than average co-channel interference. Focusing once again on the downlink, when a cell-edge user is scheduled to receive data from its serving base station, CoMP directs other base stations that are potential sources of interference not to transmit data, thus increasing SINR and data-link speeds.[47]

People looking at CoMP would naturally assume that all cell-edge uers should enjoy the benefit of this feature equally. Perhaps the thought is that if cell-edge users achieve higher data-link speeds due to improved SINR, then the capacity of the network will increase. The flaw in that argument is that when adjacent base stations are directed to mute their transmission, the stations forego the opportunity to transmit data to users, thus decreasing capacity. How does one know if the capacity gained by giving cell-edge users higher SINRs and data-link speeds are not eliminated by the effects of not transmitting data on all available Resource Blocks, which is typical of LTE networks operating at maximum capacity? Also, UPT is determined by the number of idle Resource Blocks—the higher the proportion of empty Resource Blocks, the higher the UPT. Resource Blocks are not considered idle if they are muted, so that cell-edge users in other cells can enjoy higher SINR.

My analysis uses a different approach. Instead of defining cell-edge users as those with the lowest SINR and data-link speeds, I define cell-edge users as those residing in the sector with the lowest UPT. Recall that all subscribers in the sector are assumed to have identical UPTs. Referring to the section where UPT is derived (section 2.3), the UPT of a sector is a function of the amount of unused Resource Blocks as well as the data-link speeds of all of the subscriber locations in the sector. Instead of increasing the SINR of all users in the network with the lowest data-link speeds, I focus on improving the SINR and data-link speeds of users in only the busiest sectors of the network, thus improving UPTs and capacities of those sectors. In other words, I do not attempt to improve the low data-link speeds of users in a sector with low utilization, as the UPT is likely to be good. I will use CoMP to improve the UPT of users in the most congested sectors.

You may wonder how much time and effort it took to modify the program outlined in the previous chapter to analyze a complicated feature such as CoMP, which has perplexed the industry for nearly a decade. It took less than three hours to write and debug an additional two dozen lines of code, as shown below.

47 An alternative, not to be discussed here, is when those potentially interfering base stations are directed to send copies of the same data packet to the cell-edge user.

```
%the following two lines are performed once before start of simulation
CtoIgroup = 6; %maximum C/I to enable CoMP (in dB)
[icell, busycell] = sort(subload,'descend');
%busycell is a list of cell sites sorted by the number of subscriber
%locations in descending order.

%The following lines are performed for each trial of the simulation

%Reorder active cells (ie. cells with data to be sent) with busiest cell
%first.
cellacttemp = cellactive;
cellactive = busycell(ismember(busycell, cellactive));
[Lia,Locb] = ismember(cellacttemp, cellactive);
rsub2(Locb) = rsub2;
cellon = ones(length(cellactive),1);

for j = 1:length(cellactive)
    if cellon(j)
        cplosstemp = cplossv(rsub2(j),:);
        eploss = 10.^(-cplosstemp./10);
        CtoItemp = 10*log10(eploss(cellactive(j))./eploss);
        for p = 1:j
            CtoItemp(cellactive(p)) = inf;
        end
        CtoIexempt = find(CtoItemp < CtoIgroup);
        mask(CtoIexempt) = 0;
        cellon(ismember(cellactive,CtoIexempt)) = 0;
    else  %increment busy counter for sector even though no data was sent
        numdataspeedcell(cellactive(j)) = numdataspeedcell(cellactive(j))+1;
    end
end
cellactive = cellactive(cellon == 1);
rsub2 = rsub2(cellon == 1);
```

There are a couple key items to point out. We know the busiest cell site will have the most subscriber locations, since all subscriber locations download the same amount of data, assuming that no cell site in the network is in congestion. This cell site will carry the most traffic, have the highest rate of Resource Block utilization, possess the lowest UPT and congest soonest as traffic increases. For each trial of the simulation, the first half dozen lines of the above code fragment sorts cell sites with data to transmit according to the number of subscriber locations, with the busiest cell site listed first. The busiest cell sites will enjoy the benefit of CoMP at the expense of less busy cell sites.

Starting with the busiest cell site, signal strengths measured by the subscriber are recorded and the relevant *C/I*s calculated. If the *C/I* associated with an interfering base station is less than some specified value (6 dB in the above code fragment), then the interfering base station is directed to mute its transmission if it has data scheduled to be transmitted.[48] The interfering base station will register the fact that its Resource Blocks have been used even if there was no data to be transmitted. As a result, the muted base station will have a higher utilization rate, which will reduce the UPT for its subscribers. The process is repeated for the next busiest sector with data to transmit, if that sector has not been previously directed to mute its transmission by a busier sector.

4.2.1 Impact of CoMP on KPIs

Reading research papers, you would think the benefit of CoMP can be simply stated in terms of percentage gain in capacity and data-link speeds. Perhaps this approach is preferable to stating that CoMP gains cannot be quantified because they are a function of numerous factors. Using the stochastic algorithm, I will show how various parameters affect gains in capacity and UPT. The advantage of using the stochastic algorithm is its ability to provide insights, not just calculate capacity gains.

I will use the simple four-cell site network introduced in the previous chapter to analyze CoMP. I chose two scenarios that can be considered as bookends. The first scenario is typically associated with a rural environment, where we try to maximize the downlink coverage area of an LTE network. The second scenario is more representative of a network of 5G base stations in an urban area using sub 6 GHz spectrum.

4.2.1.1 CoMP in a Rural Environment

This section will take advantage of all the data we generated for Scenario 1 from the previous chapter (η = 10, minimum-required SNR = 1). Although co-channel interference plays a role, thermal noise dominates SINR in this scenario. If co-channel interference never causes C/I to drop below 6 dB, then CoMP would have zero impact on KPIs since it would never direct any base station to mute its transmission. Remember that CoMP is triggered for a cell-edge user when A) the C/I is less than 6 dB, and B) the interfering base station carries less traffic (fewer subscriber locations); it has nothing to do with the level of the thermal noise.[49] Remind yourself of the UPT versus network traffic graph for Scenario 1 (Figure 20). Recall that the fifth percentile cell-edge UPT is basically

48 If the interfering base station has no data to be sent, then its transmitter is already muted.

49 Some may argue that it is difficult to measure the signal strength of an interfering base station if it is below the thermal noise floor. Although difficult, it is not impossible. The simulation can be modified to account for this fact.

the UPT of the busiest cell site (cell site 2), the cell site that CoMP is designed to help. Hence, when CoMP is enabled, you would expect that the fifth percentile cell-edge UPT graph should shift towards the mean UPT. (See Figure 46.)

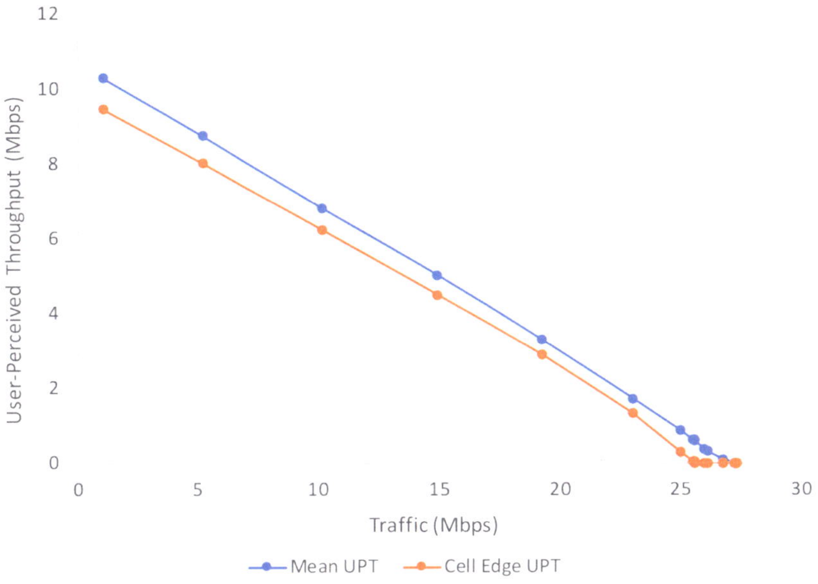

Figure 46: *Network user-perceived throughput vs traffic with CoMP enabled.*

To more easily observe the impact of CoMP, Figure 47 compares UPT vs traffic graphs with and without CoMP. As expected, CoMP improves the fifth percentile cell-edge UPT. However, the mean UPT is slightly degraded when CoMP is enabled. If one defines network capacity as the amount of traffic the network carries while maintaining a certain mean UPT, CoMP reduces network capacity. I think it is more appropriate to judge CoMP on its ability to improve cell-edge UPT, since network planners make investment decisions based on the KPIs of the busiest cell sector. If network capacity is the amount of traffic the network can carry while maintaining a fifth percentile cell-edge throughput of 2 Mbps, then it can be argued that CoMP increases network capacity from 18.5 Mbps to 21.5 Mbps, a 16% improvement. However, the mean UPT decreases from 4.1 Mbps to 2.4 Mbps, a 41% drop. Unlike other analyses of CoMP which simply state it provides x% of capacity gain, my analysis presents potential trade-offs. Is it worth the additional 16% gain in capacity if it means a drop in UPT of 41%? To further complicate the decision, CoMP is a feature that has prerequisites before it can be implemented in the LTE network. CoMP requires LTE base stations be synchronized, a practice that is not common and not inexpensive to deploy. Also, certain versions of

CoMP may require that a large percentage of the existing handsets be upgraded to support the feature before any benefits can be realized.

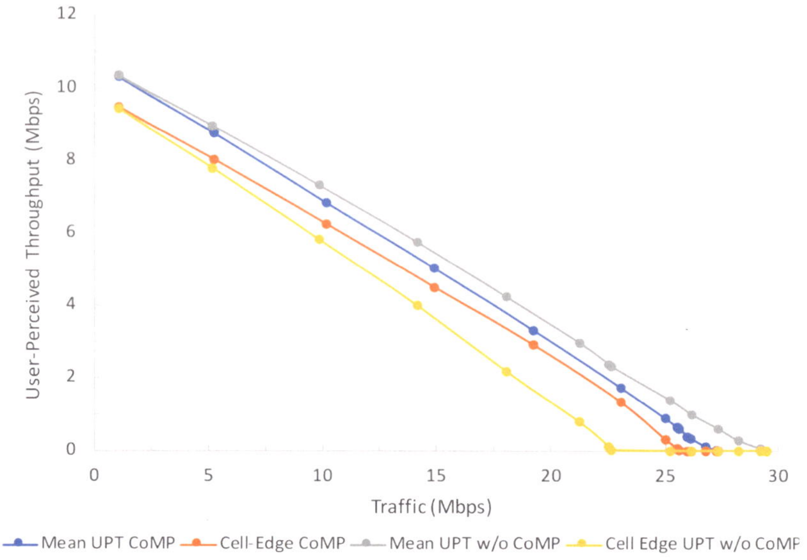

Figure 47: *Network user-perceived throughput vs traffic with and without CoMP.*

To getting a better understanding of how CoMP affects network behaviour, let's see how CoMP affects the sector traffic vs network utilization figure. Figure 26 shows the graph for Scenario 1 without CoMP and is reproduced below for your convenience.

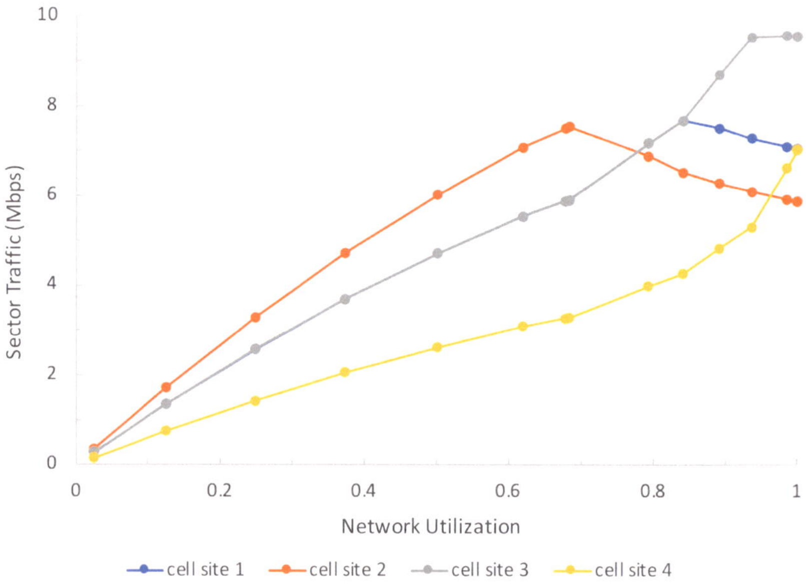

Figure 48: *Cell-site traffic versus network utilization for Scenario 1 (no CoMP).*

When network utilization surpasses 68%, the amount of traffic carried by cell site 2 decreases. Cell site 2 is in congestion (100% Resource Block utilization) and further increases in network utilization will create additional co-channel interference for subscriber locations in cell site 2, decreasing data-link speeds and capacity. With CoMP enabled, the cell-site traffic versus network utilization graph changes dramatically. (See Figure 49.)

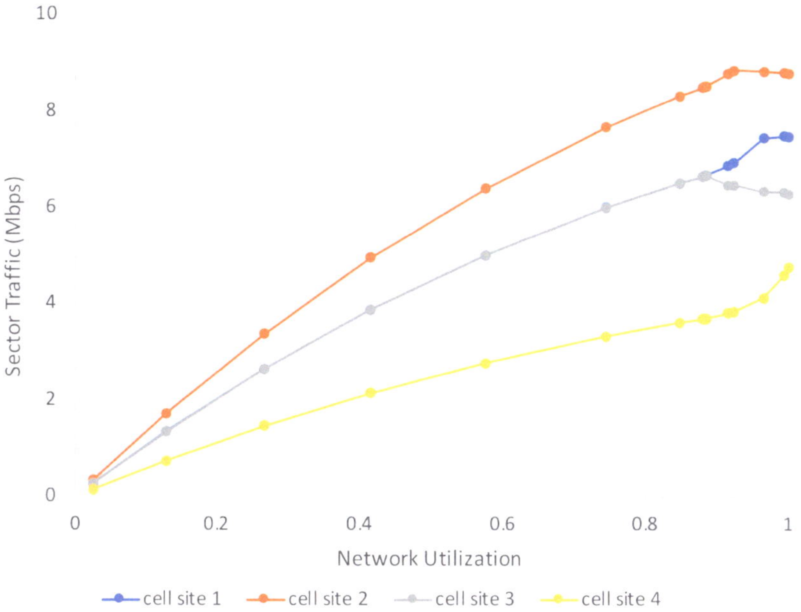

Figure 49: *Cell-site traffic versus network utilization with CoMP enabled.*

Traffic for cell site 2 does not reach congestion until network utilization surpasses 92%. In fact, cell site 1 reaches congestion first at 88% network utilization. With CoMP, the traffic for cell site 2 does not decrease appreciably after reaching congestion as base stations that interfere with cell site 2 subscribers' ability to receive data are directed to mute their transmission.

4.2.1.2 CoMP and 5G in an Urban Environment

Scenario 5—where thermal noise is significantly lower than co-channel interference at 100% utilization—is the complement to the scenario examined in the previous section. We will also eliminate the use of the CRS for idle Resource Blocks, thus mimicking 5G wireless networks in a dense urban environment. Recall from the discussion in section 3.6 about some of the anomalous behaviours we observed with this scenario, especially how in certain circumstances it appears that UPT increases with increasing traffic. The behaviour is not eliminated with CoMP, although it is somewhat muted, as seen in the figure below.

Figure 50: Network user-perceived throughput vs traffic with and without CoMP for 5G in urban environments.

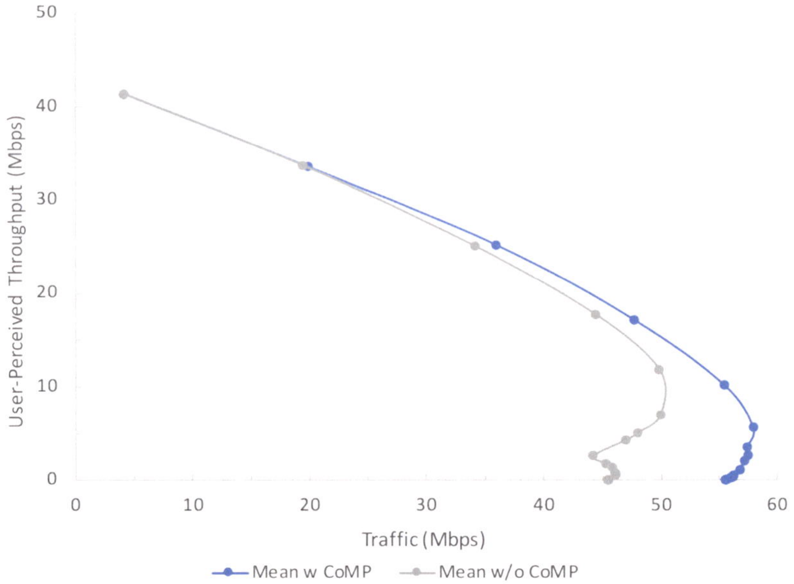

Without CoMP, it was mentioned that it was risky to operate the network beyond 38% utilization due to the devastating effect on UPT. With CoMP, it is possible to increase utilization to 56% without compromising UPT. It is possible to increase capacity with CoMP (44 Mbps to 55 Mbps) at the expense of UPT (18 Mbps to 10.25 Mbps).

Does the above discussion suggest that CoMP be deployed? This is an economic question and with existing LTE networks there may be other issues to consider, such as upgrading the base of handsets to support the feature. For a greenfield deployment, I see no harm in mandating that all handset be capable of supporting CoMP. Overall, I think CoMP is useful, but it is not the game changer that some proponents claim.

The case for Carrier Aggregation is easier to support, since it is a simpler network upgrade for the MNO and there is no need to replace handsets with CA-enabled handsets to enjoy higher UPT. The market has also spoken, since CA is universally deployed while CoMP will likely only be deployed in greenfield scenarios, if at all.

5 | Extending the Stochastic Method

When the stochastic method was first introduced in this book, we used a very simple four omni-directional cell-site network to illustrate the main features of the model. (See Figure 16.) In practice, I have never analyzed such a simple network, since typical networks have dozens of cell sites and sectors. Even this simple network provided interesting insights that no amount of intuition or thought experiment could have derived. The focus was primarily on the trade-off between traffic and UPT. This chapter will extend the method to include analysis of larger wireless networks, which will allow us to analyze the impacts of not providing ubiquitous coverage. While the goal of 100% coverage is desirable for mobile service, ubiquitous coverage is neither necessary nor always desirable for fixed service.

There will be more emphasis on providing results of analyses rather than the tool used to generate them: i.e. the stochastic algorithm examined in Chapter 3. Some results are presented in the form of fairly large and practically unreadable tables (Table 35 to Table 38). To get the maximum benefit, the reader can enter these tables into a spreadsheet and modify various parameters to see the impact on certain KPIs. If this is beyond the intent of the reader, the remainder of the book can be read without loss of comprehension.

The first major extension will be the inclusion of sectored cell sites. All macro-cellular wireless networks employ high-gain directional antennas to increase the number of sectors at a cell site from one to, typically, three

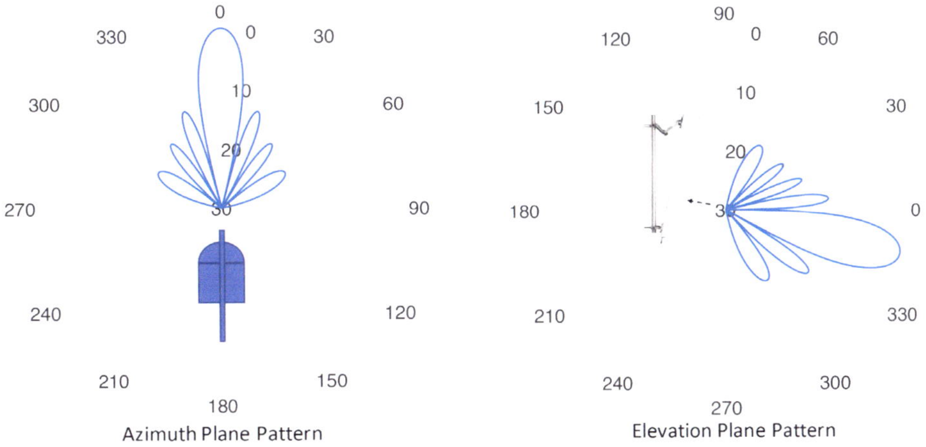

Figure 51: Antenna pattern of a directional antenna.

or more. When a mobile network operator converts a cell site from one sector to three sectors, the goal is to triple the capacity of the cell site. Depending on the characteristic of the directional antenna deployed, it is more likely that the capacity of the cell site will be at least double.[50] Figure 51 shows the antenna pattern of a typical directional antenna.

One of the main jobs of a cell-site planner is to select the most appropriate antenna and point it in the direction that will provide the most benefit to the wireless network.

In Figure 16, we used four triangles to show the locations of the four omni-directional cell sites. With sectored cell sites, arrows depict the direction of the antennas as shown in Figure 52.

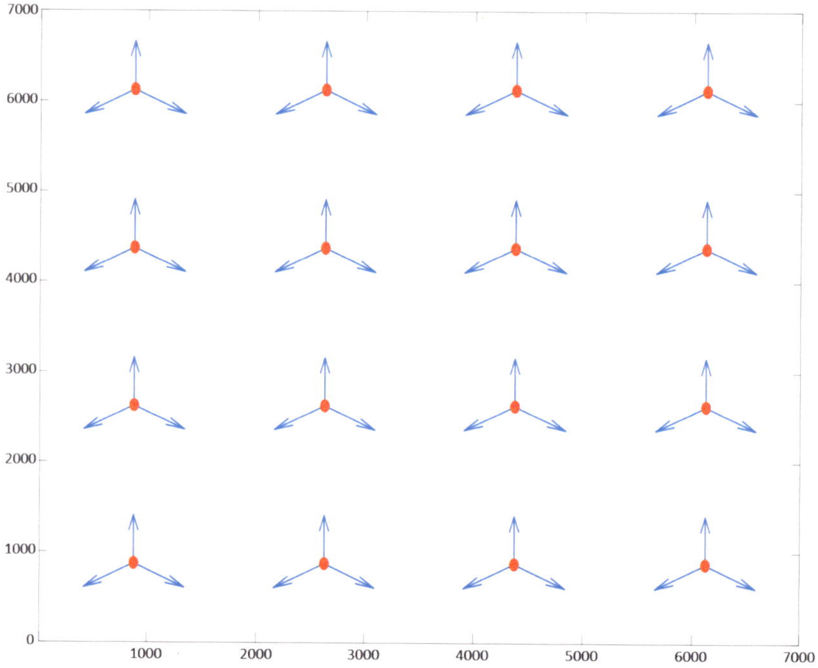

Figure 52: *A wireless network with 16 sectorized cell sites. There is a total of 3 x 16 = 48 sectors.*

This chapter is most useful if you intend to do any of the following:

- bid on spectrum in an auction
- acquire spectrum from a company selling spectrum
- acquire a company that has spectrum
- trade spectrum with another company
- answer regulator's questions
- design guidelines for the deployment of a new wireless service
- analyze the advantages and disadvantages of different spectrum bands.

50 We will see in the next chapter that there is a 20% reduction in spectral efficiency when sectorizing an omni-directional cell site into three sectors.

Key in this chapter is the development of formulas that relate how spectral efficiency, coverage and UPT are affected by various parameters. Once these formulas are established, they will be put into a spreadsheet where analysis that once took months to perform can be done instantly by altering a single spreadsheet cell.

Before I get into deriving a bunch of new formulas, I would like the reader to ponder this variation of the hypothetical scenario depicted in Chapter 1. You work for the operations department at an MNO and someone from M&A tells you that a competitor is offering 10 MHz of FDD spectrum (2 × 5 MHz) in a dense urban area where you have already fully deployed 100 MHz of FDD spectrum (2 × 50 MHz). The spectrum offered is in a lower frequency band that has not been deployed in your network. You are asked what is the value of the extra 10 MHz of FDD spectrum? Do you tell M&A that the company has exhausted its spectrum holdings and low-frequency spectrum has better propagation characteristics; therefore, M&A should pay the competitor's asking price to acquire the additional 10 MHz? In Chapter 1, it was assumed that the additional spectrum was in the same spectrum band as the existing inventory, while this chapter assumes it is in a new band where assets have not been deployed. A possible answer is given in the final section of this chapter. If you don't wish to complete this chapter, skip to the last section as the answer does not require a comprehensive understanding of all the material in the next pages.

5.1 Correlated Log-Normal Fading

Recalling Equation (31), the term '*F*' represents the long-term log-normal fading component. For a given subscriber location, there is an RF path-loss equation for every cell-site sector in the network, each containing a normal random variable *F*. In the simple four omni-directional cell-site network, each *F* is assigned a normally distributed random number independently of all the other *F*s. Every cell site is at a different location; hence, it is assumed the factors that affect long-term fading are different for each path-loss equation.

In this chapter, we are using cell sites with three sectors each. We cannot assume that *F* for each path-loss equation is independent of all other *F*s, since the path of RF propagation for a particular subscriber to each sector of the same cell site should be somewhat similar. To account for this similarity, the *F*s for each path-loss equation to a particular cell site are correlated. When the *F*s are correlated, they are still assigned different random numbers. However, if one *F* is assigned a number greater than 0, the other correlated *F*s will also be more likely to be assigned a number greater than 0.

Correlation is specified by a number between -1 and 1. In our model, each *F* is a normal random variable with the same standard deviation. If two *F*s are 100% correlated, both will be the same number. If two *F*s are 0% correlated (i.e. uncorrelated), knowing one *F* will give you no hint what the other *F* might be. If the correlation between two *F*s is -1, then both will be the same number but with opposite signs.

The amount of correlation between two cell sectors is a function of how far the cell sector antennas are apart. If the sectors are on different cell sites, then the amount of correlation between *F*s should be 0. If the sectors are on the same cell site, then the amount of correlation should be closer to 1. To capture both scenarios, I use the following function to establish the correlation between any two long-term fading component terms, based on how far the cell-site antennas are apart:

$$correl = e^{-\left({dist}/{correl\_dist}\right)} \quad (55)$$

Where *dist* is the distance between two cell-site antennas and *correl_dist* is a parameter used to establish the correlation based on *dist.* I typically set *correl_dist* to 50. Since cell sites are at least several hundred units apart, the correlation between long-term fading components from two different cell sites will be essentially zero. For example, if *dist* = 1000, then exp(-1000/50) = 2×10^{-9}.

For antennas on the same cell site, I typically set *dist* = 5. The correlation coefficient is $e^{-5/50} = 0.9$. Different values for *dist* and *correl_dist* can be used to see the effects of long-term fading correlation on key performance indicators.

It may appear that including the effects of correlation between long-term fading components is a lot of work. Actually, it isn't. The following MATLAB code can be used to establish the distance between cell sectors (matrix `ddist`).

```
for k = 1:numcells
    latb = latlongb(k,1);
    longb = latlongb(k,2);
    ddist(k,:) = sqrt((latb-latlongb(:,1)).^2+(longb-
        latlongb(:,2)).^2)+5;
    ddist(k,k) = 0;
end
```

The correlation matrix `cordist` is established using the following code:

```
correl = exp(-ddist/correldist);
```

The matrix used to produce correlated random variables is generated by the following code:

```
u = chol(correl);
```

Where `chol` is the Cholesky factorization function. Correlated fading is added by using the following line:

```
cplosst = plosssub + lognormstd*randn(size(plosssub))*u;
```

Where `plosssub` is the RF path loss without the long-term fading component, `lognormstd` is the standard deviation of the long-term fading component and `cplosst` is the RF path loss including the long-term fading component.

5.2 Key Performance Indicators

In order to compare the effectiveness of various network strategies, we need to measure three key performance indicators: network capacity, coverage and user-perceived throughput. There are different ways to specify the capacity of a network. One method used by vendors is to specify the amount of traffic a network can carry when every Resource Block is carrying data (i.e. 100% Resource Block utilization). In most cases, when every Resource Block is carrying data traffic, the network is at its maximum[51]. However, we know from using Equation (11) that the user-perceived throughput is 0 Mbps for every subscriber in the network. Clearly, the Marketing Department might be concerned if its advertising campaign is based on having the fastest network.

A different definition for the capacity of the network is the amount of traffic it can carry while meeting some particular UPT metric. The mobile network operator may focus on the average or cell-edge UPT meeting a specified number. For example, a wireless network operator may bid on a contract that states that 95% of the subscribers must see data throughputs of at least “x” Mbps during the busy hour.[52]

Another alternative is to specify the amount of traffic a network can carry when 30% of the Resource Blocks are carrying data. In a network with a large number of cell-site sectors, when the average sector is at 30% utilization, there will be sectors approaching 100% utilization due to the uneven distribution of traffic. I fortuitously found this statement on page 28 of the FCC REPORT AND ORDER AND FURTHER NOTICE OF

51 An exception was illustrated when co-channel interference dominates thermal noise and reference symbols were absent. (See Figure 38.)

52 The 5th percentile cell-edge user perceived throughput must be greater than “x” Mbps.

PROPOSED RULEMAKING 16-89, July 14, 2016, which is the Spectrum Frontiers Proposal to enable flexible-use 5G services at mmWave frequency bands in the United States.

> *"...both base stations and user equipment could be entirely silent much of the time; terrestrial operators report that, in current deployments, network loading rarely exceeds 30 percent."*

The first thing I do when analyzing a new wireless network is to produce the same table I introduced when describing the development of the stochastic method with the four omni-directional cell site simple example. (See Table 21: Template for displaying network and cell-site KPIs.) With the current network we are analyzing, I will show only the network KPIs, since displaying the individual cell-sector KPIs would consume too many pages.

Table 24: *Network KPIs.*

subs selected	1	2	4	6	8	10	12	14	16	18
trials	120000	60000	40000	45000	48000	40000	40000	30000	30000	30000
spec eff (bps/Hz)	0.04	0.07	0.14	0.19	0.23	0.27	0.30	0.32	0.35	0.36
se active RBs (bps/Hz)	1.85	1.78	1.64	1.52	1.40	1.30	1.21	1.13	1.06	1.00
traffic (Mbps)	18.54	35.65	65.52	90.71	111.47	129.08	143.18	155.84	165.69	173.89
user-perceived throughput (Mbps)	18.45	17.37	15.30	13.53	11.95	10.59	9.36	8.32	7.38	6.54
SINR (dB)	15.71	15.27	14.44	13.69	12.99	12.37	11.78	11.25	10.73	10.28
C/I (dB)	15.71	15.27	14.44	13.69	12.99	12.37	11.78	11.25	10.73	10.28
SNR (dB)	59.37	59.37	59.37	59.37	59.37	59.37	59.37	59.37	59.37	59.37
10th cell edge (Mbps)	15.47	14.44	12.59	10.71	9.05	7.77	6.62	5.65	4.76	4.01
5th cell edge (Mbps)	14.41	13.41	11.60	10.04	8.63	7.39	6.45	5.49	4.69	3.68
utilization	0.02	0.04	0.08	0.12	0.17	0.21	0.25	0.29	0.33	0.36

Table 25: *Network KPIs continued.*

subs selected	20	25	30	35	50	75	100	200	300	500
trials	10000	10000	10000	10000	10000	10000	10000	10000	10000	10000
spec eff (bps/Hz)	0.38	0.41	0.42	0.43	0.45	0.50	0.52	0.53	0.53	0.53
se active RBs (bps/Hz)	0.94	0.83	0.75	0.69	0.60	0.55	0.53	0.53	0.53	0.53
traffic (Mbps)	181.75	195.76	202.72	207.93	217.38	239.06	248.72	255.25	255.25	255.25
user-perceived throughput (Mbps)	5.83	4.41	3.40	2.67	1.43	0.45	0.13	0.00	0.00	0.00
SINR (dB)	9.86	8.94	8.20	7.63	6.67	6.10	5.80	5.71	5.71	5.71
C/I (dB)	9.86	8.94	8.20	7.63	6.67	6.10	5.80	5.71	5.71	5.71
SNR (dB)	59.37	59.37	59.37	59.37	59.37	59.37	59.37	59.37	59.37	59.37
10th cell edge (Mbps)	3.41	2.03	1.11	0.46	0.00	0.00	0.00	0.00	0.00	0.00
5th cell edge (Mbps)	2.97	1.73	0.94	0.00	0.00	0.00	0.00	0.00	0.00	0.00
utilization	0.40	0.49	0.56	0.63	0.76	0.91	0.97	1.00	1.00	1.00

Both Table 24 and Table 25 show the KPIs for the wireless network in Figure 52, in the scenario in which co-channel interference is the major factor in determining network performance. The adjusted thermal noise factor η is set to an extremely low value (1×10^{-5}), resulting in a very high SNR (59.37 dB). Hence, the SINR is basically the same as *C/I*, which you can see in the tables. These tables assume that 10 MHz of LTE is deployed and both base station and mobile receiver support 2x2 MIMO. Data-link speeds as a function of SINR are specified in Figure 4. After I generate the table, I make some graphs. The most important one is the User-perceived throughput vs traffic graph. (See Figure 53.)

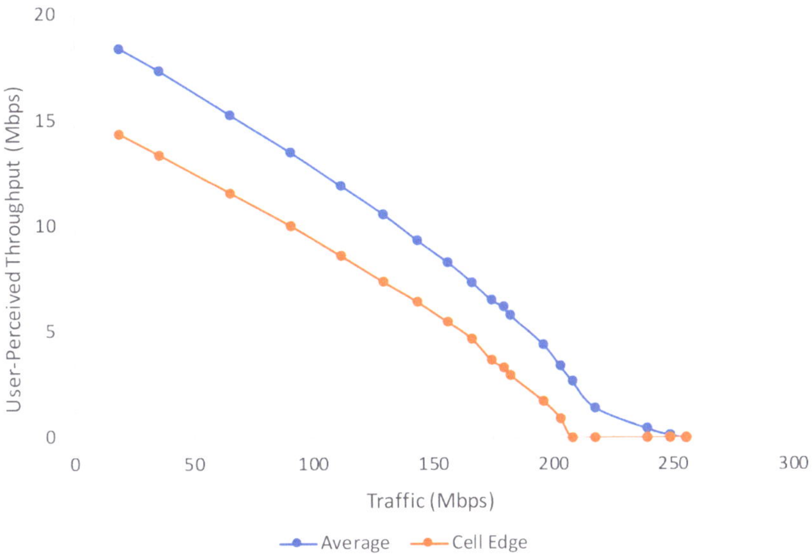

Figure 53: *User-perceived throughput vs traffic for a 16 cell-site network.*

A related graph is shown below; this is part of the graph of Figure 53 that extends below what I considered the capacity of the network.

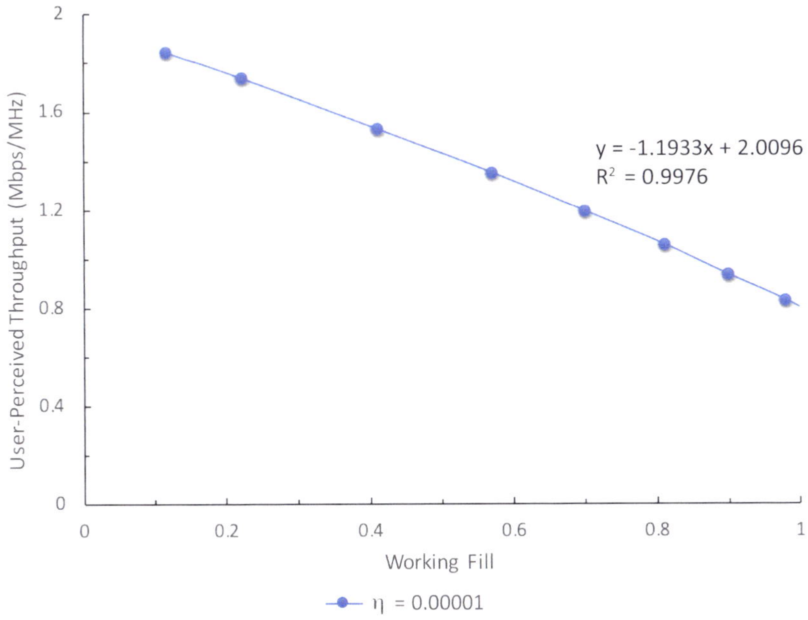

Figure 54: *User-perceived throughput (Mbps/MHz) vs working fill.*

In this case, I consider that the network is at 100% capacity when 30% of the Resource Blocks are carrying data (utilization = 0.30). Referring to Table 24, 30% utilization occurs between 14 and 16 subscribers selected per simulation. The user-perceived throughput KPI is also normalized in Figure 54, so that it is measured in Mbps/MHz. The reason for this will become apparent when we compare networks with different spectrum bandwidth.

You will also notice that Figure 54 has a new term called "Working Fill." I use 100% working fill to indicate the level of traffic at which the customer's Quality of Experience is considered acceptable, which in this case occurs at 30% network utilization.

Referring to Figure 54, when the network is at 100% working fill, the average user-perceived throughput is 0.8 Mbps/MHz. When the network is carrying less traffic (i.e. working fill is less than 100%), one of the benefits is higher UPTs. The relationship between user-perceived throughput and working fill is quite linear. Embedded in Figure 54 is an equation that gives the linear equation between the normalized UPT and working fill, which I will simplify with the following equation:

$$u = -1.2f + 2. \tag{56}$$

Where u is the normalized user-perceived throughput and f is the working fill. Note that the y-intercept of 2 is the UPT per MHz at 0% working fill. This equation will be useful in quantifying trade-offs between traffic and Quality of Experience.

As I mentioned previously, I am defining network capacity as when Resource Block utilization is 30%. Unfortunately, Table 24 only has Resource Block utilization for 29% and 32%. Linear interpolation is used to derive the KPIs for 30% utilization, which are summarized in the following table.

Coverage	100%
Spectral efficiency (bps/Hz)	0.33
Spectral efficiency active RBs (bps/Hz)	1.11
Traffic (Mbps)	159.25
User-perceived throughput (Mbps/MHz)	0.8
SINR (dB)	11.07
C/I (dB)	11.07
SNR (dB)	59.37

Table 26: *KPIs at 30% Resource Block utilization.*

Recall the discussion in section 2.1 where the conversion from bps/Hz to GB/month/MHz was discussed. The spectral efficiency in Table 26 could also be expressed as 33 GB/month/MHz.

We like to examine the effects of thermal noise on network KPIs when co-channel interference is not so dominant. This occurs when the cell sites are further apart, or when the subscriber is located deep inside a building, or a higher frequency band is being analyzed, etc. This is achieved by running simulations with different values for the parameter η. The only problem occurs when choosing values for η so large that the value for SINR for certain subscriber locations becomes unacceptably low. Recalling section 3.3, we defined whether a subscriber location was in coverage by comparing the strength of the desired signal with η. In a real network, the mobile network operator cannot control which mobile subscribers access the network based on downlink signal strength.[53] In one sense, we are performing a thought experiment in which we pretend that we can exclude subscribers when their SNR does not meet a specified minimum value. Simulations will inform us how the minimum-required signal strength will affect the KPIs. For example, by increasing the minimum-required signal strength, the proportion of subscriber locations meeting the minimum-required signal strength (the coverage percentage) will decrease.

5.3 Factors that Impact Coverage

When η is no longer insignificant compared to co-channel interference, we need to determine whether the signal strength of the desired signal (typically the strongest base-station signal) is sufficiently high that, even in the absence of co-channel interference, the signal can be received and demodulated (i.e. the receiver is in coverage). To determine whether the signal strength of the desired signal is sufficient at a particular subscriber location, we need:

- The value of η to be known or be set to a certain value
- The minimum-required signal-to-noise ratio to be known or be set to a certain value.

The minimum-required signal strength (minSS) can be expressed in two ways:

$$minSS = \eta \times SNR_{min}, \tag{57}$$

or,

$$minSS\ (dB) = \eta\ (dB) + SNR_{min}(dB). \tag{58}$$

53 Recall that the uplink can impose limitations on the maximum-allowable RF path loss, thus affecting the minimum-required downlink signal strength where the mobile can establish communications.

Higher values of η and higher minimum SNR requirements will increase the minimum-required signal strength. Higher minimum-required signal strength will exclude more subscriber locations, thus reducing coverage. Note that the minimum-required signal strength is what determines coverage and any combination of η and minimum-required SNR that produces the same minimum-required signal strength will have the same coverage. Recall the simple four-cell network in Chapter 3, where we examined five scenarios. (See Table 23 for a summary.) One parameter was identical for all five scenarios—the minimum-required signal strength (minSS) —which was set to 10. Of the 80 initial random subscriber locations, 69 met that criteria. Each scenario had different values for η and SNR_{min}, which impacted various KPIs, except for coverage which is only a function of minSS.

These three parameters quite often appear together, much like The Three Stooges. Perhaps it will help the reader to think of the other parameters when one of them is referred to. It may further help the reader to associate each of the above parameters with one of The Three Stooges. Just a thought.

Even though the following two tables have different values for η and minimum-required SNR, the minimum-required signal strength for both tables is identical, thus the coverage for both is identical.

Table 27: *Coverage vs minimum-required signal strength when minimum-required SNR = 0.3. Coverage is obtained through simulations.*

η	Minimum-Required SNR	Minimum-Required Signal Strength (minSS)	Coverage
0.033	0.3	0.01	100.0%
0.100	0.3	0.03	99.9%
0.333	0.3	0.1	99.4%
1.000	0.3	0.3	95.9%
3.333	0.3	1	82.0%
10.000	0.3	3	62.2%
33.333	0.3	10	41.6%
100.000	0.3	30	25.3%
333.333	0.3	100	14.8%
1000.000	0.3	300	8.4%

Table 28: *Coverage vs minimum-required signal strength when minimum-required SNR = 1. Coverage is obtained through simulations.*

η	Minimum-Required SNR	Minimum-Required Signal Strength (minSS)	Coverage
0.01	1	0.01	100.0%
0.03	1	0.03	99.9%
0.1	1	0.1	99.4%
0.3	1	0.3	95.9%
1	1	1	82.0%
3	1	3	62.2%
10	1	10	41.6%
30	1	30	25.3%
100	1	100	14.8%
300	1	300	8.4%

Below is a graph of Tables 27 and 28.

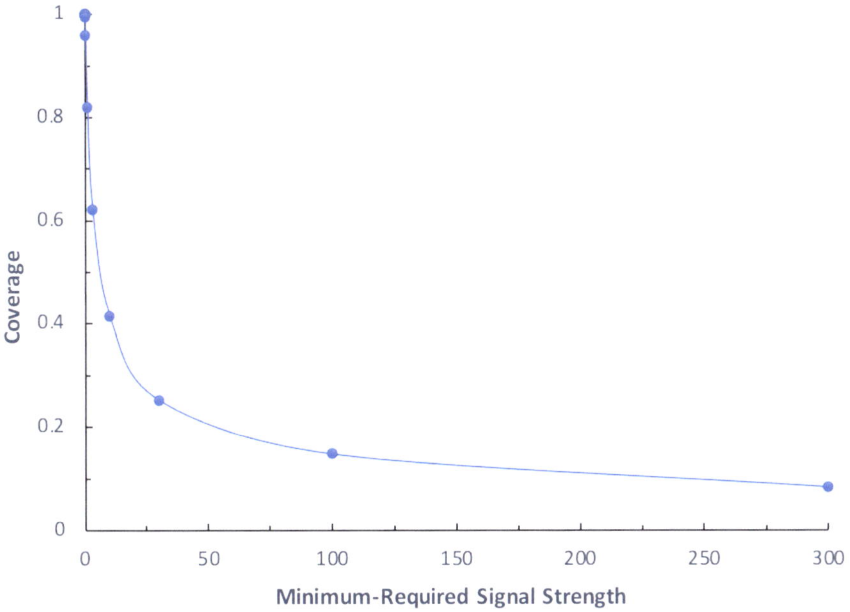

Figure 55: *Coverage vs minimum-required signal strength.*

Although illustrative, some transformations will provide further insights.

Minimum-Required Signal Strength (minSS)	$10\log_{10}(\text{minSS})$	Coverage	$-\log_{10}(\text{Coverage})$
0.00001	-50.00	1.000	0.000
0.0001	-40.00	1.000	0.000
0.001	-30.00	1.000	0.000
0.01	-20.00	1.000	0.000
0.03	-15.23	0.999	0.000
0.1	-10.00	0.994	0.003
0.3	-5.23	0.959	0.018
1	0.00	0.820	0.086
3	4.77	0.622	0.206
10	10.00	0.416	0.381
30	14.77	0.253	0.597
100	20.00	0.148	0.830
300	24.77	0.084	1.078

Table 29: *Transformation of coverage and minimum-required signal strength.*

Below is the graph of the highlighted columns of Table 29.

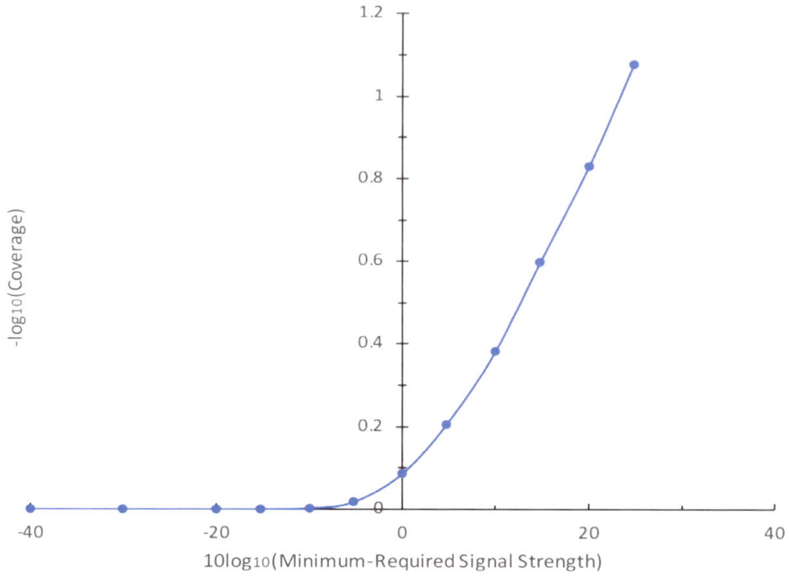

Figure 56: *$-log_{10}$(coverage) vs $10log_{10}$(minimum-required signal strength).*

For those acquainted with the Shannon-Hartley equation (1), the graph in Figure 56 should look familiar. The following table is of the following equation:

$$\beta = \left[log_{10}\left(\frac{minSS}{2.25}+1\right)\right]/2. \tag{59}$$

Note that β is a temporary variable that will be used to derive another formula that will give the predicted coverage for a given minSS. Once the equation for the predicted coverage is determined, you will never see β or its formula (shown above) again.

Table 30: *Finding an equation that will predict coverage.*

Minimum-Required Signal Strength (minSS)	$10log_{10}$(minSS)	Actual Coverage	$-log_{10}$(Coverage)	β Equation (59)	Predicted Coverage $10^{-\beta}$
0.00001	-50.00	1.000	0.000	0.000	1.000
0.0001	-40.00	1.000	0.000	0.000	1.000
0.001	-30.00	1.000	0.000	0.000	1.000
0.01	-20.00	1.000	0.000	0.001	0.998
0.03	-15.23	0.999	0.000	0.003	0.993
0.1	-10.00	0.994	0.003	0.009	0.978
0.3	-5.23	0.959	0.018	0.027	0.939
1	0.00	0.820	0.086	0.080	0.832
3	4.77	0.622	0.206	0.184	0.655
10	10.00	0.416	0.381	0.368	0.429
30	14.77	0.253	0.597	0.578	0.264
100	20.00	0.148	0.830	0.829	0.148
300	24.77	0.084	1.078	1.064	0.086

A comparison of the 4th and 5th column of Table 30 is shown in the figure below.

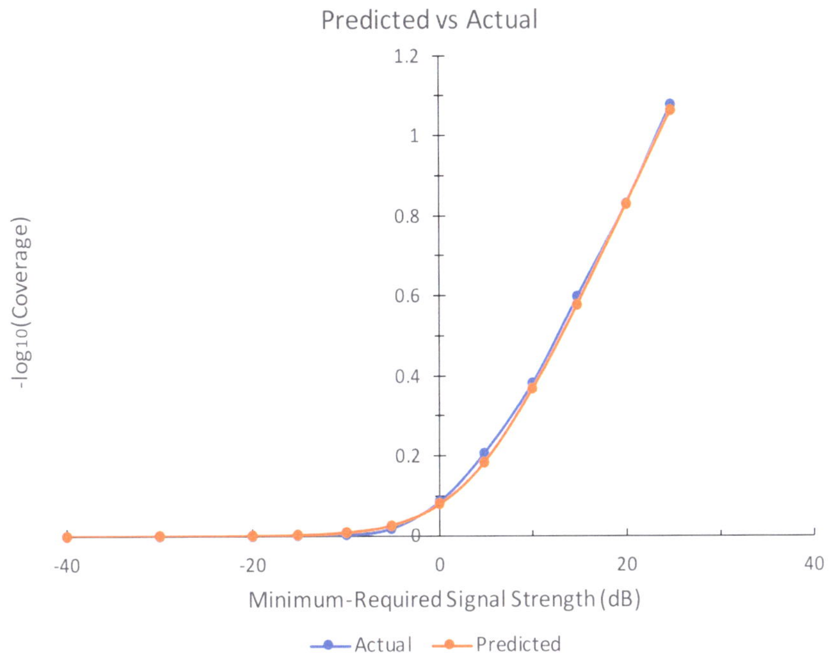

Figure 57: *Predicted vs actual. Actual coverage is from the results of simulations.*

One can also inspect the 3rd column of Table 30 (Actual Coverage) with the last column (Predicted Coverage) to see that both provide approximately the same value for coverage, given the minimum-required signal strength.

The predicted coverage formula can be simplified:

$$predicted\ coverage = 10^{-\left[log_{10}\left(\frac{minSS}{2.25}+1\right)\right]/2},$$
$$= \left(\frac{minSS}{2.25}+1\right)^{-1/2}. \qquad (60)$$

Hence, once the minimum-required signal strength is calculated, the coverage can be predicted by using Equation (60). Note that Equation (60) is only applicable for the network that we are analyzing in Figure 52. Equation (60) needs to be re-calculated for each particular network configuration.

5.4 Factors that Impact Capacity

Figure 53 is the typical graph when η is set to an extremely low value. The following graph shows what happens when η is set to a much higher value (η = 1 and the minimum-required SNR is also set to 1 or 0 dB).

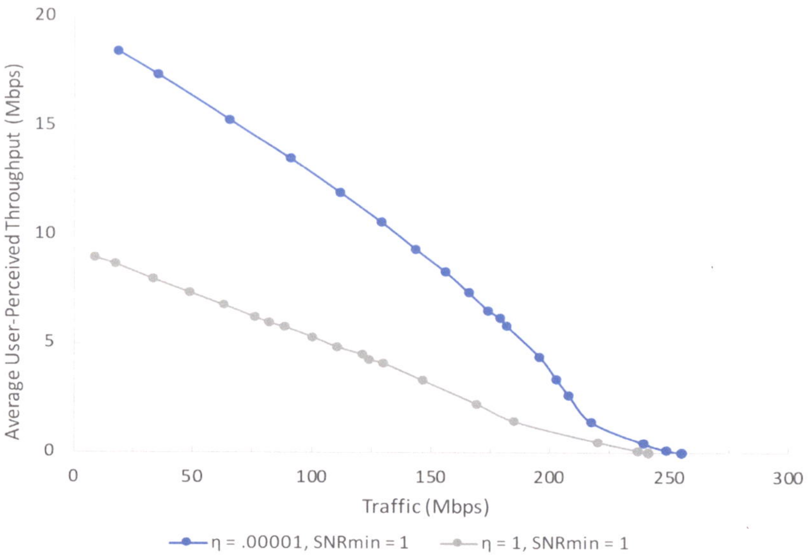

Figure 58: *UPT vs traffic for different levels of adjusted thermal noise.*

When η is set to a higher number, both the amount of traffic and the UPT are lower. The following table quantifies the differences when both scenarios are operating at 100% working fill, which in our case means that Resource Block utilization is at 30%.

Scenario	η=0.00001	η=1
Coverage	100%	82%
Spectral efficiency (bps/Hz)	0.33	0.215
Spectral efficiency active RBs (bps/Hz)	1.11	0.719
Traffic (Mbps)	159.25	103.39
User-perceived throughput (Mbps/MHz)	0.8	0.52
SINR (dB)	11.07	8.17
C/I (dB)	11.07	12.51
SNR (dB)	59.37	12.20

Table 31: *KPIs at 30% Resource Block utilization and minimum-required SNR = 1.*

Notable points from going from η = 0.00001 to η = 1:

- Coverage has decreased from 100% to 82%
- Traffic/spectral efficiency has decreased 35%
- User-perceived throughput has also dropped 35%[54]
- Average co-channel interference has gotten better with the reduction in coverage (average *C/I* has improved from 11.07 dB to 12.51 dB)
- Average SINR has decreased from 11.07 dB to 8.17 dB due to the much higher thermal-noise level.

Note that an increase in η from 0.00001 to 1 can represent a change in morphology from urban to rural. According to the model, the spectral efficiency should be lower for rural cell sites, opposite what is sometimes predicted by the heuristic method. (See section 1.2.) The mystery deepens.

One of the more important facets is that the UPT vs working fill graph is linear. The slope and y-intercept will change for different values of η and minimum-required SNR. (See figure below.)

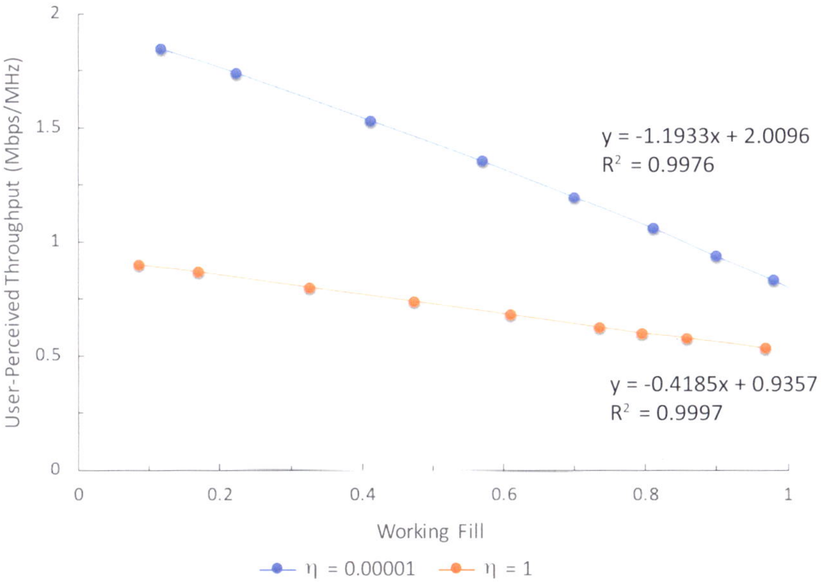

Figure 59: *Normalized user-perceived throughput vs working fill (minimum-required SNR = 1).*

54 The fact that the decreases in traffic and user-perceived throughput are identical is somewhat intuitive but not immediately obvious. One could analyze Equation (11) and state that both traffic and user-perceived throughput are a linear function of the capacity of the network ρ_c. Increases in η would cause ρ_c to decrease by a certain percentage, causing both traffic and user-perceived throughput to decrease by the same amount. Or one could appeal to the analysis in the chapter analyzing Carrier Aggregation, where doubling spectrum bandwidth doubles both capacity and UPT. Regardless, it is nice when the data confirms the intuition. There are some benefits when two KPIs change by the same amount, such as reduced complexity in the analysis, which the reader will later appreciate.

It was previously mentioned that minimum-required SNR may not be known, hence it could be made variable to establish its effect on network KPIs. For mobile networks, I will analyze two scenarios. In the first, the minimum-required SNR will be set to a low value of 0.3 or -5.23 dB, while in the second it is set to 1 or 0 dB.

Now, we have a decision to make. We can either analyze the network using η or we can use the minimum-required signal strength (minSS). In one sense, they are not all that different, since η and minSS only differ by the minimum-required SNR, which is set to a specific constant. (See Equations (57) and (58) or see Table 27 and Table 28.) The argument for using η is that SINR depends on η, as can be seen in Equation (49), then directly affects the data-link speed as determined by either the Shannon-Hartley equation (1) or Figure 4. The argument for using minSS is that its value determines which subscriber locations are to be included in the analysis (see previous section), and that will have a large impact on *C/I* and, ultimately, the SINR. At one time, I would have favoured using η as the dependent variable in most of my analysis, but now I prefer minSS due to its importance in selecting subscriber locations, which has been the focus of my user-centric approach. As stated above, the three terms minSS, η, and SNR_{min} quite often appear together and it is helpful to know how each are related to the other two.

The goal now is to extend Table 24 and Table 25 to show how η , the minimum-required SNR and minimum-required signal strength affect three important KPIs:

- traffic at 100% working fill
- user-perceived throughput at 100% working fill
- user-perceived throughput at 0% working fill.

We have the basic user-perceived throughput vs working fill (UPT vs WF) graph for the scenario when co-channel interference dominates η (see Figure 54) and now we need to see how the graph is affected when η becomes more significant. (See, for example, Figure 59.) We have seen that the reduction in traffic and UPT at 100% working fill are identical. (See footnote 54.) This means we only need to determine the reduction in UPT at 100% working fill and 0% working fill. This will give the slope and the y-intercept of the lines in Figure 59. Once we have the linear equation for the UPT vs WF graph, we can estimate the average UPT for any working fill between 0% to 100%.

Table 32 is the drop in average UPT for the scenario, where the minimum-required SNR is 0.3 or -5.23 dB. Note that UPT drop = (1 – Relative UPT). The figure below shows the last two columns of Table 32.

η	Minimum-Required SNR	minSS	Relative UPT @ 0% WF	Relative UPT @ 100% WF	UPT drop @ 0% WF	UPT drop @ 100% WF
0.000033	0.3	0.00001	100.0%	100.0%	0.0%	0.0%
0.00033	0.3	0.0001	99.8%	99.9%	0.2%	0.1%
0.0033	0.3	0.001	98.4%	98.7%	1.6%	1.3%
0.033	0.3	0.01	86.3%	91.7%	13.7%	8.3%
0.100	0.3	0.03	71.2%	81.7%	28.8%	18.3%
0.333	0.3	0.10	48.8%	63.3%	51.2%	36.7%
1.000	0.3	0.30	33.3%	47.1%	66.7%	52.9%
3.333	0.3	1.00	25.0%	37.7%	75.0%	62.3%
10.000	0.3	3.00	20.3%	33.1%	79.7%	66.9%
33.333	0.3	10.00	19.1%	31.1%	80.9%	68.9%
100.000	0.3	30.00	19.2%	31.2%	80.8%	68.8%
333.333	0.3	100.00	18.5%	30.4%	81.5%	69.6%
1000.000	0.3	300.00	17.9%	30.1%	82.1%	69.9%

Table 32: *Average user-perceived throughput drops at various minimum-required signal strengths (minimum-required SNR = 0.3).*

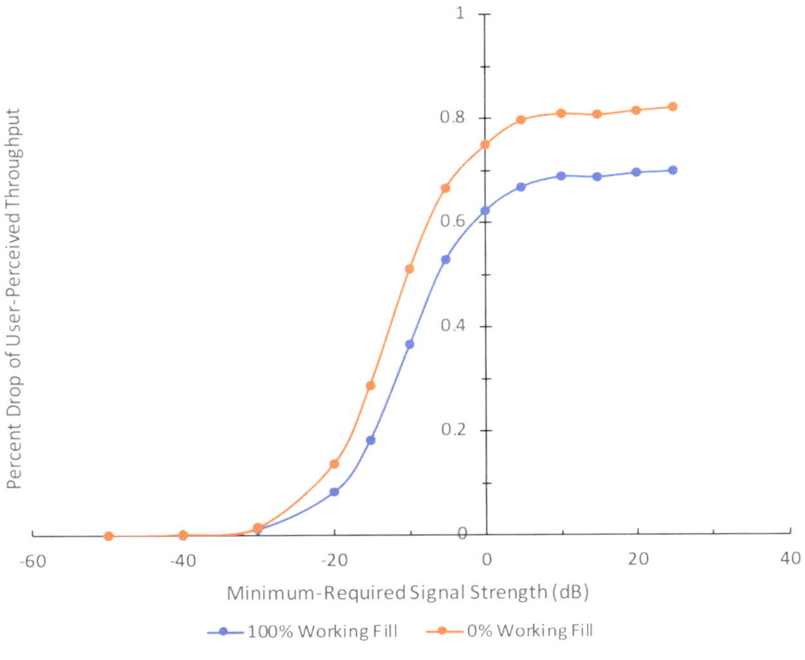

Figure 60: *Percentage drop in average user-perceived throughput versus minimum-required signal strength (minimum-required SNR = 0.3).*

Those who are familiar with functions may suspect that the curves in the above graph resemble the logistic function:

$$upt\ drop = \frac{c}{(1 + ab^x)}. \tag{61}$$

Where x is minSS in dB. A very simple algorithm is used to determine the parameters a, b and c of the logistic function used to quantify the drop in average UPT. I will describe the algorithm by estimating the parameters for the drop in average UPT in the case of 0% working fill.

ESTIMATING *c*

The estimate for 'c' is obtained by determining the maximum value of the graph. In the above figure, it appears that the maximum value of the average UPT drop is approximately 81%.

ESTIMATING *a*

The estimate for 'a' is obtained by setting x to 0 in Equation (61), and solving for 'a' once the average UPT drop for $x = 0$ is determined. Note that when $x = 0$, b^x is equal to 1. In the case for the 0% Working Fill curve, the average UPT drop is 75%. Hence:

$$75 = \frac{81}{1 + a}$$

$$a = \frac{81}{75} - 1 = 0.08\,.$$

ESTIMATING *b*

Once c and a have been determined, b is estimated by picking another point on the curve and solving for b. For example, the average UPT drop at $x = -20$ dB is 13.7%. Hence:

$$13.7 = \frac{81}{1 + 0.08b^{-20}}$$

$$b = 0.81\,.$$

Therefore, the average UPT drop for 0% working fill is given by:

$$UPT\ drop = \frac{81}{1 + 0.08(0.81)^x} \tag{62}$$

Figure 61 shows how well Equation (62) and the data in the 4th column (Relative UPT @ 0% WF) of Table 32 fit. Remember that Relative UPT = 1 - UPT drop.

Table 33 summarizes the parameters of the four UPT logistic curves of interest.

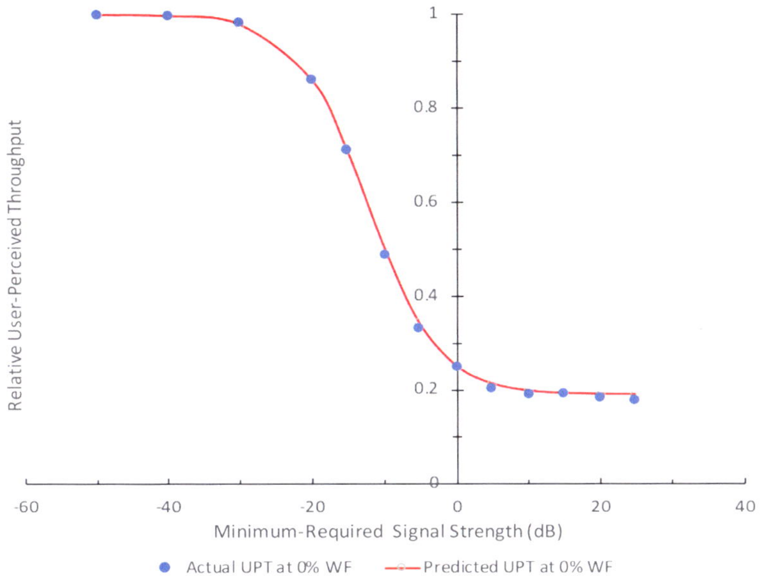

Figure 61: *Predicted curve vs actual UPT data points at 0% working fill (minimum-required SNR = 0.3).*

Working Fill	0%	100%	0%	100%
Minimum-Required SNR	0.3 (or -5.23 dB)	0.3 (or -5.23 dB)	1 (or 0 dB)	1 (or 0 dB)
a	0.08	0.11	0.17	0.086
b	0.81	0.81	0.81	0.77
c	81	69.5	62	38.3

Table 33: *Parameters for UPT logistic curves.*

Remember, the logistics curve gives the reduction or drop in UPT. For completeness, the figure below shows the relative UPT (which is 1 – UPT drop) for the four cases examined in the above table.

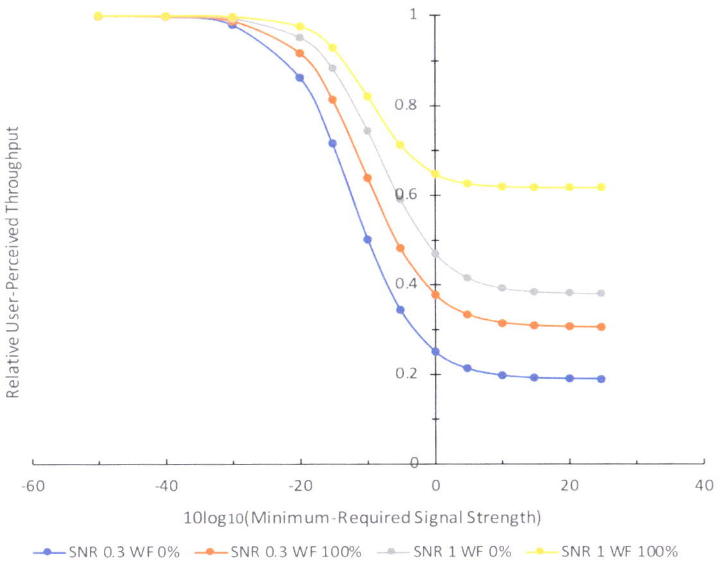

Figure 62: *Relative user-perceived throughput vs minimum-required signal strength.*

Recall Equation (56), which gives the equation for the line in Figure 54. This line is only applicable for extremely low values of η. Figure 59 displays an additional line for the case when η = 1 and minimum-required SNR = 1. We now have the tools to derive the appropriate linear equation for any combination of η, minSS and minimum-required SNR. Starting with Equation (56) we can generalize it:

$$u = -1.2f + 2 = mf + b\,.$$

Note that f = 0 at working fill = 0% and f = 1 at working fill = 100%. Thus, setting f = 0, we get:

$$u(0) \equiv u_0 = m \times 0 + b = b = 2\,.$$

Thus, the y-intercept b is just the UPT at working fill = 0% (no traffic). Setting f = 1, we get:

$$u(1) \equiv u_1 = m \times 1 + u_0 = m + u_0\,.$$
$$m = u_1 - u_0\,.$$

For later calculations, $u_1 \equiv u(1) = -1.2 + 2 = 0.8$ Mbps/MHz.

Now we derive the linear equation for UPT vs working fill for non-trivial η.

$$u' = m'f + b'.$$
$$u'_0 = b'. \qquad \text{Set } f = 0.$$
$$u'_1 = m' + u'_0\,. \qquad \text{Set } f = 1.$$
$$m' = u'_1 - u'_0\,.$$

Using the scenario where η = 1 and minimum-required SNR is 1 (or 0 dB), we need to find the UPT at working fill = 0% (i.e. u'_0) and the UPT at working fill = 100% (i.e. u'_1). Recall that x = 10*$\log_{10}$(minimum required Signal Strength) = 10*$\log_{10}$(η × minimum required SNR) = 10$\log_{10}$(1 × 1) = 0 dB. Using Equation (61) and Table 33:

(63)
$$u'_0 = \left(1 - \frac{c/100}{(1 + ab^x)}\right)u_0\,,$$

$$u'_0 = \left(1 - \frac{62/100}{(1 + 0.17 \cdot 0.81^0)}\right)u_0 = 0.47u_o = 0.94\,.$$

(64)
$$u'_1 = \left(1 - \frac{c/100}{(1 + ab^x)}\right)u_1\,,$$

$$u'_1 = \left(1 - \frac{38.3/100}{(1 + 0.086 \cdot 0.77^0)}\right)u_1 = 0.65u_1 = 0.52\,.$$

$$m' = u'_1 - u'_0 = 0.52 - 0.94 = -0.42\,.$$

$$u' = m'f + b' = -0.42f + 0.94\,.$$

The last equation matches the lower line in Figure 59.

5.5 Spectrum Analysis

I have now inundated you with a large amount of information. However, all these equations can be entered into a spreadsheet no larger than this page, with an approximate file size of 20 KB. Although there is a learning curve in becoming proficient in the method outlined in this chapter, it's nothing compared to the heuristic method described in the first chapter. That involves analyzing tens of thousands of cell sites and generating spreadsheets on the order of 50 MB that take months to modify. The spreadsheet I will develop can analyze new scenarios in a matter of milliseconds.

Detractors of the stochastic method will complain that many of the model's parameters are not realistic. For example, it does not include the impact of the signalling/control channel, rectangular grid spacings are rarely used in actual networks, the geographical distribution of the subscriber locations is too uniform, etc. However, the business questions typically posed are comparisons between two or more possible scenarios, and it is unlikely that these idealistic assumptions will favor one scenario over another.

I will present a situation some of you may encounter if you work for a mobile network operator, or as an analyst in the financial industry responsible for valuating telecom stocks, or your responsibilities may involve underwriting telecom bonds. Imagine there is a scheduled government auction of spectrum licences involving both low-band spectrum (635 MHz) and high-band spectrum (2,165 MHz). One question could be: which is more valuable: 10 MHz of low-band spectrum or 15 MHz of high-band spectrum? Do you believe low-band spectrum is always better? Or do you tell interested stakeholders, "Low-band spectrum is passé, and mid-band spectrum is the new beachfront property since it is possible to acquire larger bandwidths." We will answer the question by analyzing the effects of frequency and bandwidth on a number of KPIs (coverage, capacity and UPT) in several scenarios (rural, urban and dense urban). Those in the industry realize these can be multi-billion-dollar questions. The return on the investment of good analysis is nearly immeasurable.

As a reminder, all equations and graphs we have derived in this chapter assume that the base station and mobile device supports 2×2 MIMO.

5.5.1 Rural Analysis

In order to avoid too many generalities, we will analyze a rural area of 2,500 km^2. (For example, an area 50 km by 50 km). We will assume there are 10 subscribers per square kilometre, therefore there are 25,000 subscribers in the study area. One of the important input parameters is the total monthly network traffic. To generate traffic, we will assume that the mobile network operator aims to have 20% market share and each subscriber downloads 2 GB/month. The target monthly traffic is simply 25,000 × 0.2 × 2 = 10,000 GB/month. The cost of providing service will be based on the number of cell sites required to provide both coverage and capacity.

We will assume that the mobile network operator has a target to cover 95% of the area under study. We can reduce by 5% the number of subscribers and total monthly network traffic to account for the less than 100% coverage. The table below captures the market parameters of this analysis:

Scenario	10 MHz of 635 MHz	15 MHz of 2135 MHz
Size of study area (km^2)	2,500	2,500
Subscribers/km^2	10	10
Total subs in study area	25,000	25,000
Target share assuming 100% coverage	20%	20%
Target number of subs	5,000 × 0.95	5,000 × 0.95
Average usage per sub (GB/month)	2	2
Network traffic (GB/month)	10,000 × 0.95	10,000 × 0.95
Total number of cell sites	?	?

Table 34: *Input parameters of rural analysis.*

We need to determine η and the minimum-required signal strength (minSS) for both cases. From the previous discussion, η and minSS differ by the minimum-required SNR, as shown by Equations (57) and (58). I will perform two analyses where I assume the minimum-required SNR is either 0.3 (-5.23 dB) or 1 (0 dB).

5.5.1.1 Coverage Analysis - Rural

The target coverage is 95%. Using Equation (60), we can calculate the largest minimum-required signal strength (minSS) to achieve 95% coverage.

$$0.95 = \left(\frac{minSS}{2.25} + 1\right)^{-1/2},$$

$$minSS = 2.25\left(\frac{1}{0.95^2} - 1\right) = 0.243 \; or \; -6.143 \; dB\,.$$

Using Equation (58) we can calculate η:

$$\eta\;(dB) = minSS\;(dB) - SNR_{min}(dB),$$
$$\eta\;(dB) = -6.143 \;\; for\; SNR_{min}(dB) = 0\; dB$$
$$\eta\;(dB) = -0.914 \;\; for\; SNR_{min}(dB) = -5.23\; dB$$

Recall from Equation (48) the definition of η. η (dB) is the exponent of Equation (48).

$$\eta\ (dB) = rnf - Transmit\ Power - Gain + Misc\ Loss + 45.5 + (35.46 - 1.1h_m)\log f_c + 0.7h_m + C + 10\gamma\log\tau - 13.82\log h_b\,. \quad (65)$$

Equation (65) gives η (dB), but what we really want is the scale factor τ. The network we are analyzing (see Figure 52) has an inter-cell site distance of 1,750 meters. If we keep that distance (i.e. we set the scale factor τ to 1), we would find that η (dB) = -24.22 dB and not the value we are seeking (η (dB) = -6.143 dB or -0.914 dB, depending on SNR_{min}). Hence, we must re-write the above equation so we can solve for τ:

$$10\gamma\log\tau = \eta\ (dB) - rnf + Transmit\ Power + Gain - Misc\ Loss - 45.5 - (35.46 - 1.1h_m)\log f_c - 0.7h_m - C + 13.82\log h_b\,, \quad (66)$$

$$\tau = 10^{(\eta\ (dB) - rnf + Transmit\ Power + Gain - Misc\ Loss - 45.5 - (35.46 - 1.1h_m)\log f_c - 0.7h_m - C + 13.82\log h_b)/10\gamma} \quad (67)$$

Due to the number of terms, it helps to use the table below:

Scenario	SNR_{min} = 0 dB		SNR_{min} = -5.23 dB	
	Low band	High band	Low band	High band
Noise factor, η (dB)	-6.143	-6.143	-0.914	-0.914
Bandwidth (MHz)	10	15	10	15
Receiver noise floor (dBm / 10 MHz)	-97	-97	-97	-97
Frequency (MHz)	635	2165	635	2165
Transmit Power (W)	40	40	40	40
Transmit power (dBm / 10 MHz)	46.02	44.26	46.02	44.26
Mobile antenna gain (dB)	0	0	0	0
Base antenna gain (dB)	17.7	17.7	17.7	17.7
Total gain (dB)	17.7	17.7	17.7	17.7
In-building losses (dB)	4.36	13.85	4.36	13.85
Miscellaneous losses (dB)	4.36	0.00	4.36	0.00
Constant 1	45.5	45.5	45.5	45.5
Mobile-antenna height (m)	1.5	1.5	1.5	1.5
$(35.46 - 1.1h_m)\log f_c$	94.76	112.77	94.76	112.77
C	0	0	0	0
Base-antenna height (m)	70	70	70	70
$13.82\log h_b$	25.50	25.50	25.50	25.50
$10\gamma\log\tau$ (sum of the above shaded cells)	34.41	18.99	39.63	24.22
Path-loss exponent, 10**γ**	40	40	40	40
Scale factor, **τ** (see eqtn (67))	7.25	2.98	9.79	4.03
Nominal inter-cell site distance (m)	1750	1750	1750	1750
Inter-cell site distance (m)	12681.7	5222.8	17135.4	7057
Cell site coverage area (km^2)	160.8	27.27	293.6	49.8
Number of cell sites required for 2500 km^2	15.5	91.7	8.5	50.2

Table 35: *Inter-cell site distance and number of cell sites calculations for 95% coverage rural scenario.*

The terms inside the shaded cells are the terms that are either added or subtracted in Equations (66) and (67). The other numbers are required to calculate the terms in the shaded cells. All the terms in the above table should be familiar, but pay attention to the miscellaneous losses term. In the Low-band scenario, that term include the in-building losses. 3GPP uses the following formula to calculate in-building penetration loss for different frequency bands:

(68) $$loss = 7.7409 \times \ln(f_{MHz}) - 45.597\,.$$

At 635 MHz, in-building penetration loss is 4.36 dB while at 2165 MHz it is 13.85 dB. I've included the in-building loss for the low band to demonstrate its ability to penetrate walls. The high losses at 2,165 MHz make it uneconomical to provide in-building coverage for this rural scenario. Thus, this is not an apples-to-apples comparison. The low-band network provides in-building coverage while the high-band network does not. Even so, the low-band network has far fewer cell sites, as shown in the table above.

Note that since each of the four scenarios has the exact same coverage, if we ran a simulation of each of the four scenarios, the exact same subscribers would have been selected. The difference between the scenarios would have been the scale, i.e., the distance between various points, whether they are cell sites or subscriber locations. The study area in Figure 52 is no longer 7,000 m by 7,000 meters, but some distance given by $\tau \times 7000$ m. Note that there are half as many cell sites for the scenarios where $SNR_{min} = 0.3 = -5.23$ dB. The subscriber locations can receive signals that are 5.23 dB weaker than in the scenario, where the minimum-required SNR is 1 or 0 dB, which means the subscribers can be further away from the cell sites (35% further away).

5.5.1.2 Capacity Analysis - Rural

Are the number of cell sites sufficient to support the amount of traffic generated by the subscribers? We know from Table 26 that the spectral efficiency when co-channel interference dominates thermal noise is 0.33 bps/Hz or 33 GB/month/MHz. We now need to know the spectral efficiency when co-channel interference is not so dominant. Spectral efficiency will decrease since co-channel interference is still present, whilst the additional thermal noise will further reduce the SINR. Recall that the drop in spectral efficiency will be equal to the drop in UPT at 100% working fill (denoted as u_1). We also need to determine the drop in UPT at 0% working fill (denoted as u_0). Instead of making you read more text, it is perhaps easier to present the information as a table:

Although there are a lot of items in Table 36, it is not difficult to put all this information into a spreadsheet. (The data was actually copied directly from a spreadsheet). In fact, I encourage the reader to do so and then start modifying the parameters to see how it would affect the results. Each modification is equivalent to potentially many months of work using the heuristic method or many hours (even days or weeks) of computer time performing Monte Carlo simulations, if the programs developed by industry or academia are used.

Table 36: Capacity calculations for rural scenario.

	Scenario	SNR_{min} = 0 dB		SNR_{min} = -5.23 dB	
		Low band	High band	Low band	High band
a	Coverage	95%	95%	95%	95%
b	Bandwidth (MHz)	10	15	10	15
c	Number of cell sites (Table 35)	15.5	91.7	8.5	50.2
d	Number of sectors (d = 3 × c)	46.6	275.0	25.5	150.6
e	Size of study area (km²)	2500	2500	2500	2500
f	Subs/km²	10	10	10	10
g	Total subs in study area (g = e × f)	25,000	25,000	25,000	25,000
h	Target share assuming 100% coverage	20%	20%	20%	20%
i	Share (i = a × h)	19%	19%	19%	19%
j	Number of subs (j = i × g)	4750	4750	4750	4750
k	Subs per sector (k = j ÷ d)	101.9	17.3	186.0	31.5
l	Average usage per sub (GB/month)	2	2	2	2
m	Total monthly traffic (GB/month) (m = j × l)	9500	9500	9500	9500
n	Logistic parameter 'a' working fill = 0% (Table 33)	0.17	0.17	0.08	0.08
o	Logistic parameter 'b' working fill = 0% (Table 33)	0.81	0.81	0.81	0.81
p	Logistic parameter 'c' working fill = 0% (Table 33)	0.620	0.62	0.820	0.82
q	u(0) UPT $\eta \approx 0$ working fill = 0% (56)	2	2	2	2
r	Logistic parameter 'a' working fill = 100% (Table 33)	0.086	0.086	0.11	0.11
s	Logistic parameter 'b' working fill = 100% (Table 33)	0.77	0.77	0.81	0.81
t	Logistic parameter 'c' working fill = 100% (Table 33)	0.383	0.383	0.695	0.695
u	u(1) UPT $\eta \approx 0$ working fill = 100% (56)	0.800	0.800	0.800	0.800
v	u'(0) UPT η>>0 working fill = 0% ((63) using n, o, p)	1.23	1.23	0.73	0.73
w	u'(1) UPT η>>0 working fill = 100%((64) using r, s, t)	0.59	0.59	0.40	0.40
x	m' (x = w – v)	-0.65	-0.65	-0.33	-0.33
y	Spectral efficiency (bps/Hz) (Table 26)	0.33	0.33	0.33	0.33
z	Spectral efficiency (GB/month/MHz) (z = 100 × y)	33	33	33	33
aa	Traffic adjustment (1 - (61) using r, s, t)	0.73	0.73	0.50	0.50
ab	Traffic capacity at 100% working fill (GB/Month/MHz) (ab = z × aa)	24.15	24.15	16.63	16.63
ac	Traffic per sector (GB/Month/MHz) (ac = m ÷ d ÷ b)	20.37	2.30	37.19	4.21
ad	Working fill rate (ad = ac ÷ ab)	84.3%	9.5%	223.6%	25.3%
ae	Average data speed (Mbps/MHz) (ae = x × ad + v)	0.69	1.17	0.00	0.65
af	Avg user perceived throughput (Mbps) (af = ae × b)	6.87	17.59	-0.01	9.72

You may notice that one cell is shaded. The network using low-band spectrum and minimum-required SNR set at -5.23 dB does not have sufficient capacity to support the traffic demand. The network is capacity constrained. In all the other scenarios, there is sufficient capacity, hence, we could say the networks are coverage constrained. The way I designed Table 35 and Table 36 assumed that the network would be coverage constrained. In most situations you may not know which constraint applies. So, despite my advice to produce a spreadsheet using the above tables as a template, perhaps the reader should wait to see how the tables are modified to handle the capacity-constrained scenario.

5.5.1.3 User-Perceived Throughput Analysis - Rural

Inspecting Table 35 and Table 36, it is obvious that 10 MHz of low-band spectrum is more desirable than 15 MHz of high-band spectrum, due to the significantly lower number of cell sites and better in-building coverage. However, there is one advantage with the 15 MHz option: higher average UPT. Focusing on the scenario in which the minimum-required SNR is 1 (or 0 dB), the UPT for the high-band network is 17.89 Mbps vs 6.87 Mbps for the low-band network. This is because there are so many high-band cell sites needed to achieve 95% coverage, the required working fill is only 9.5%[55]. Also, the high-band network has 15 MHz of spectrum versus only 10 MHz for the low-band network[56].

If you think the trade-offs still favour the 15 MHz high-band option, perform a quick financial analysis. Let's assume a rural cell site costs $500,000 to construct, while the equipment costs $50,000 per sector. Since the cell sites in Figure 52 have three sectors, the approximate cost of a rural cell site is $650,000. In row k of Table 36, the number of subscribers per sector for the high-band network is 17.1 for SNR_{min} = 0 dB and 31.5 for SNR_{min} = -5.23 dB, or equivalently 51.3 subscribers per cell site for SNR_{min} = 0 dB and 94.5 for SNR_{min} = -5.23 dB. If the average subscriber pays $50/month, then the average cell site generates 51.3 × 50 = $2,565/month or $30,780/year for SNR_{min} = 0 dB. For SNR_{min} = -5.23 dB, the monthly and annual revenues are $4,725 and $56,700, respectively. From these revenue figures, you must subtract the cell site OPEX costs, which can range from $3,000 to $4,000 per month. Clearly, there is no business case for deploying wireless service on high-band spectrum in rural areas.

55 See Figure 53 for a reminder of the trade-off between traffic and user-perceived throughput.
56 Refer to (1) for the relationship between bandwidth and data-link speeds.

5.5.2 Urban Analysis

In a capacity-constrained network, the above tables are slightly modified. The number of cell sites in the study area is a function of the inter-cell site spacing, using the following equation:

$$\#\ cell\ site = \frac{area}{inter\ site\ distance^2}\ . \qquad (69)$$

The total capacity of the network is the product of the number of cell-site sectors,[57] the spectral efficiency and the amount of spectrum bandwidth:

$$capacity = 3 \times \#\ cell\ sites \times spec\ eff \times bandwidth\ ,$$

$$capacity = 3 \times \left(\frac{area}{inter\ site\ distance^2}\right) \times spec\ eff \times bandwidth\ . \qquad (70)$$

Since the traffic is known, we can set capacity to the known traffic and use the above equation to solve for the inter-cell site distance. Unfortunately, the spectral efficiency is also a function of inter-site distance, so that method will not work in all cases[58]. The solution is to set the inter-cell site distance to a certain value, calculate the number of cell sites using Equation (69), calculate the spectral efficiency for that cell site spacing and use Equation (70) to find the capacity of the network. If the capacity of the network is greater than the expected traffic, a larger cell-site spacing is chosen and the process is repeated. Obviously, if the capacity of the network is less than the expected traffic, a smaller cell site spacing is chosen and the process is repeated. Fortunately, most popular spreadsheet programs have a feature that implements a goal-seeking algorithm; use this to find the inter-cell site spacing that will achieve 100% working fill for a given level of traffic.[59]

As in the previous section, I could write many pages of texts or put the information in a chart. Table 37 is a comparison of 10 MHz of low-band spectrum versus 15 MHz of high-band spectrum in a 50 km × 50 km urban environment.

57 Remember, the number of sectors in our example network is three times the number of cell sites.
58 If the inter-cell site spacing is small enough, we can use the spectral efficiency number when co-channel interference dominates thermal noise (i.e. 0.33 bps/Hz or 33 GB/month).
59 Recall that the capacity of the network is the amount of traffic at 100% working fill.

Table 37: *Coverage/capacity analysis for urban scenario.*

Scenario		Low band	High band	Low band	High band
a	Minimum-required SNR	1	1	0.3	0.3
b	Minimum-required SNR (dB)	0.00	0.00	-5.23	-5.23
c	Nominal inter-cell site distance (m)	1750	1750	1750	1750
d	Inter-cell site distance (m)	1573.0	1675.8	1572.6	1592.6
e	Scale factor, τ (e = d / c)	0.90	0.96	0.90	0.91
f	Bandwidth (MHz)	10	15	10	15
g	Receiver noise floor (dBm / 10 MHz)	-97	-97	-97	-97
h	Frequency (MHz)	635	2165	635	2165
i	Transmit power (W)	40	40	40	40
j	Transmit power (dBm / 10 MHz)	46.02	44.26	46.02	44.26
k	Mobile-antenna gain (dB)	0	0	0	0
l	Base-antenna gain (dB)	17.7	17.7	17.7	17.7
m	Total gain (dB)	17.7	17.7	17.7	17.7
n	In-building losses (dB)	4.36	13.85	4.36	13.85
o	Miscellaneous losses (dB)	4.36	13.85	4.36	13.85
p	Constant 1	45.5	45.5	45.5	45.5
q	Mobile-antenna height (m)	1.5	1.5	1.5	1.5
r	$(35.46 - 1.1h_m)\log f_c$	94.76	112.77	94.76	112.77
s	C	3	3	3	3
t	Base-antenna height (m)	35	35	35	35
u	$13.82\log h_b$	21.34	21.34	21.34	21.34
v	Path-loss exponent, 10γ	40	40	40	40
w	$\gamma\log\tau$ ($v \times \log_{10}(e)$)	-1.85	-0.75	-1.86	-1.64
x	Noise factor, η (dB) (x = g - j - m + o + p + r + s - u + w)	-35.240	-4.875	-35.245	-5.759
y	Minimum-required signal strength (dB)	-35.240	-4.875	-40.474	-10.988
z	Minimum-required signal strength	0.000	0.325	0.000	0.080
aa	Coverage	100%	93%	100%	98%
ab	Number of cell sites ($ab = ad \times 1e6 / d^2$)	1010.4	890.2	1010.9	985.6
ac	Number of sectors (ac = 3 × ab)	3031.1	2670.6	3032.7	2956.9
ad	Size of study area (km^2)	2500	2500	2500	2500
ae	Subs/sq. km	2000	2000	2000	2000
af	Total subs in study area (af = ad × ae)	5,000,000	5,000,000	5,000,000	5,000,000
ag	Target share assuming 100% coverage	10.0%	10.0%	10.0%	10.0%
ah	Share (ah = aa × ag)	10.0%	9.3%	10.0%	9.8%
ai	Number of subs (ai = ah × af)	499,967	467,339	499,990	491,378
aj	Subs per sector (aj = ai / ac)	164.9	175.0	164.9	166.2
ak	Average usage per sub (GB/month)	2	2	2	2
al	Total monthly traffic (GB/month) (al = ai × ak)	999,934	934,679	999,980	982,756
am	Logistic parameter 'a' working fill = 0%	0.17	0.17	0.08	0.08
an	Logistic parameter 'b' working fill = 0%	0.81	0.81	0.81	0.81
ao	Logistic parameter 'c' working fill = 0%	0.620	0.62	0.820	0.82
ap	u(0) UPT $\eta \approx 0$ working fill = 0%	2	2	2	2
aq	Logistic parameter 'a' working fill = 100%	0.086	0.086	0.11	0.11
ar	Logistic parameter 'b' working fill = 100%	0.77	0.77	0.81	0.81
as	Logistic parameter 'c' working fill = 100%	0.383	0.383	0.695	0.695
at	u(1) UPT $\eta \approx 0$ negligible working fill = 100%	0.800	0.800	0.800	0.800

at	u(1) UPT η ≈ 0 negligible working fill = 100%	0.800	0.800	0.800	0.800
au	u'(0) UPT η » 0 working fill = 0%	2.00	1.16	2.00	1.09
av	u'(1) UPT η » 0 working fill = 100%	0.80	0.57	0.80	0.54
aw	m' (aw = av - au)	-1.20	-0.59	-1.20	-0.56
ax	Spectral efficiency (bps/Hz)	0.33	0.33	0.33	0.33
ay	Spectral efficiency (GB/month/MHz)	33	33	33	33
az	Traffic adjustment	1.00	0.71	1.00	0.67
ba	Traffic capacity at 100% working fill (GB / Month / MHz) (ba = ay × ac)	32.99	23.33	32.96	22.15
bb	Traffic per sector (GB / Month / MHz) (bb = al / ac / f)	32.99	23.33	32.97	22.16
bc	Working fill rate (WFR) (bc = bb / ba)	100.0%	100.0%	100.0%	100.0%
bd	Average data speed (at 100% WFR) (Mbps/MHz) (bd = aw × bc + au)	0.80	0.57	0.80	0.54
be	Avg user perceived throughput (Mbps) (be = bd × f)	8.00	8.49	7.98	8.05

There are a few points I would like to highlight:

- In both low-band and high-band cases, in-building penetration losses have been included.
- The number of cell sites in all cases are similar. The network with 15 MHz of high-band spectrum and minimum-required SNR set to 1 has the fewest cell sites, although it has the least amount of coverage (93%). Network operators always like to plan their network with the intention of running it at 100% working fill, so it is not unlikely that they will sacrifice some coverage. A lower in-building penetration assumption can be used to increase coverage.
- The differences between the low-band cases where in one case $SNR_{min} = 1$ and the other $SNR_{min} = 0.3$ are negligible.
- The correction factor C in row 's' is set to 3 dB since we are analyzing an urban environment.

Although somewhat tedious, the reader should enter this table in a spreadsheet using the formulas described in this chapter. Using a feature such as goal seek, one should attempt to find the inter-cell site distance to meet any number of KPIs (working fill, average user-perceived throughput, coverage) for different network parameters (subscriber density, spectrum frequency, spectrum bandwidth, in-building penetration loss assumptions, market share, monthly subscriber data usage, radio power, antenna height, etc.). You will gain far more insight by doing this than reading text from a book.

5.5.3 Dense Urban

It is possible in certain cities for the population density to exceed 10,000 people per km^2. Alternatively, subscriber usage could grow far beyond the 2 GB/month assumed thus far in this chapter. In either case, what really matters in determining cell-site density is the traffic density in GB/month/km^2. Rather than produce another large table such as the one for the urban morphology, we can assume that smaller inter-cell site distance will cause the scale factor τ to decrease, causing $\gamma \log\tau$ to decrease, resulting in η to be a fairly small number. A small value for η means that SINR will essentially equal *C/I* (equation turns into (41)). When network performance is strictly limited by co-channel interference, network capacity is no longer a function of the frequency of the spectrum but of the bandwidth of the spectrum. Therefore, 15 MHz of high-band spectrum will carry 50% more traffic than 10 MHz of low-band spectrum, as well as have an average UPT that is 50% greater. Alternatively, 10 MHz of low-band spectrum requires 50% more cell sites than 15 MHz of high-band spectrum for the same amount of capacity.

5.5.4 Spectrum Valuation Example

Spectrum valuation is the most important task in the preparation phase before bidding begins in a spectrum auction. The table below outlines the process of determining the value of 2 × 5 MHz of spectrum for the Urban Scenario (2,000 subs/km^2). I will make the assumption that the operator has 2 × 10 MHz of high-band spectrum and will determine the value of the extra 5 MHz of spectrum by making a comparison with a network with 2 × 15 MHz of spectrum in the same band. Once again, I will present the data in a chart, although this time only the most important parameters and KPIs will be shown.

Table 38: *Urban high-band spectrum analysis.*

Scenario		10 MHz	15 MHz	10 MHz	15 MHz
a	Minimum-required SNR	1	1	0.3	0.3
b	Minimum-required SNR (dB)	0	0	-5.23	-5.23
d	Inter-cell site distance (m)	1432.0	1675.8	1413.5	1592.6
f	Bandwidth (MHz)	10	15	10	15
h	Frequency (MHz)	2165	2165	2165	2165
aa	Coverage	98%	93%	99%	98%
ab	Number of cell sites (ab = ad × 1e6 / d^2)	1219.1	890.2	1251.2	985.6
ac	Number of sectors (ac = 3 × ab)	3657.4	2670.7	3753.7	2956.9
ad	Size of study area (km^2)	2500	2500	2500	2500
ae	Subs/sq. km	2000	2000	2000	2000
af	Total subs in study area (af = ad × ae)	5,000,000	5,000,000	5,000,000	5,000,000
ag	Target share assuming 100% coverage	10.0%	10.0%	10.0%	10.0%
ah	Share (ah = aa × ag)	9.8%	9.3%	9.9%	9.8%
ai	Number of subs (ai = ah × af)	487,621	467,342	496,379	491,378
aj	Subs per sector (aj = ai / ac)	133.3	175.0	132.2	166.2
ak	Average usage per sub (GB/month)	2	2	2	2
al	Total monthly traffic (GB/month) (al = ai × ak)	975,243	934,683	992,757	982,756

ax	Spectral efficiency (bps/Hz)	0.33	0.33	0.33	0.33
ay	Spectral efficiency (GB/month/MHz)	33	33	33	33
az	Traffic adjustment	0.81	0.71	0.80	0.67
ba	Traffic capacity at 100% working fill (GB / Month / MHz) (ba = ay × ac)	26.66	23.33	26.45	22.15
bb	Traffic per sector (GB / Month / MHz) (bb = al / ac / f)	26.66	23.33	26.45	22.16
bc	Working fill rate (WFR) (bc = bb / ba)	100.0%	100.0%	100.0%	100.0%
bd	Average data speed (at 100% WFR) (Mbps/MHz) (bd = aw × bc + au)	0.65	0.57	0.64	0.54
be	Avg. user-perceived throughput (Mbps) (be = bd × f)	6.46	8.49	6.41	8.05
bf	Radio and antenna costs	$ 50,000	$ 50,000	$ 50,000	$ 50,000
bg	Cell site radio/Ant costs (bf = 3 × bf)	$ 150,000	$ 150,000	$ 150,000	$ 150,000
bh	Tower costs	$ 350,000	$ 350,000	$ 350,000	$ 350,000
bi	Total base-station cost (bi = bg + bh)	$ 500,000	$ 500,000	$ 500,000	$ 500,000
bj	Network cost (bj = ab × bi)	$609,574K	$445,112K	$625,616K	$492,810K

The most obvious consequence of having more spectrum is the reduction in the number of cell sites. When the minimum-required SNR is set to 1, it is possible to reduce the number of cell sites by 27% (1 – 890.2/1219.1). When SNR_{min} = 0.3, the reduction is 20%. In both cases, the reduction is less than 33%, which would be the expected reduction if all network configurations have the same spectral efficiency. Deploying fewer cell sites means the inter-cell site distance is larger, which results in a lower spectral efficiency (see rows az and ba) and a slight reduction in coverage.

Another advantage of additional spectrum is an increase in the average UPT. Once again, the gain in average UPT is less than the 50% increase that would be expected with a 50% boost in spectrum bandwidth, due to the decrease in spectral efficiency.

People are always interested in the value of spectrum expressed in $/MHz/POP. It is relatively easy to calculate the value of the additional 2 × 5 MHz of spectrum. Using the example when SNR_{min} = 1, the 27% reduction in the number of cell sites results in a 27% CAPEX savings. With the radio and base-station costs assumed in this example, approximately $164,462,000 is saved. Dividing this number by the population (5,000,000) and the amount of spectrum (2 × 5 = 10 MHz) results in a spectrum value of $3.29/MHz/POP. Of course, this number is very sensitive to the assumptions regarding traffic, population and equipment costs. In fact, it is not too difficult to develop a scenario that will meet any $/MHz/POP one can imagine. The goal is to obtain insight, although this tool can be used/misused to provide data to justify any particular spectrum acquisition strategy.

5.6 Signal-to-Noise Ratio Analysis

If you took the time to implement the spreadsheets in this chapter, I would not be surprised if you found them a little bit unwieldy to use. It would probably take some time playing with them to get a good understanding of what is happening. When I first developed these spreadsheets, although I was confident in the results they were producing, I didn't have a good idea how it was doing it. I fed this black box with numbers, and it spat out answers like magic. This section attempts to provide more insights of how the various parameters such η, minSS and SNR_{min} interact to produce the KPIs for the various network parameter settings.

In this chapter, we used rather low numbers for the minimum-required SNR (SNR_{min} = 1 or 0.3), which is indicative of a system where the uplink and downlink maximum allowable RF path losses are similar. What happens if the links are not balanced? As mentioned previously, the output power of the base-station radio is typically 100 times more powerful than the transmitter in the smartphone. To deliver adequate uplink data speeds, a lower maximum-allowable path loss must be assumed, which means the minimum-required SNR for the downlink could be higher by 10 dB to 20 dB. In other words, the uplink governs what appears to be a downlink parameter.

For the examples in this chapter where SNR_{min} is set to a value less than or equal to 1 (or 0 dB), higher η and minimum-required signal strength results in lower spectral efficiency. If one recalls that the drop in traffic is equal to the drop in UPT at 100% working fill, we can use Figure 62 to see the effects of the minimum-required signal strength when SNR_{min} is set to values less than or equal to 1. When SNR_{min} is greater than 3 (or 4.8 dB), the spectral efficiency increases with higher η and minimum-required signal strength, as shown Figure 63.

Figure 63: *Traffic vs minimum-required signal strength.*

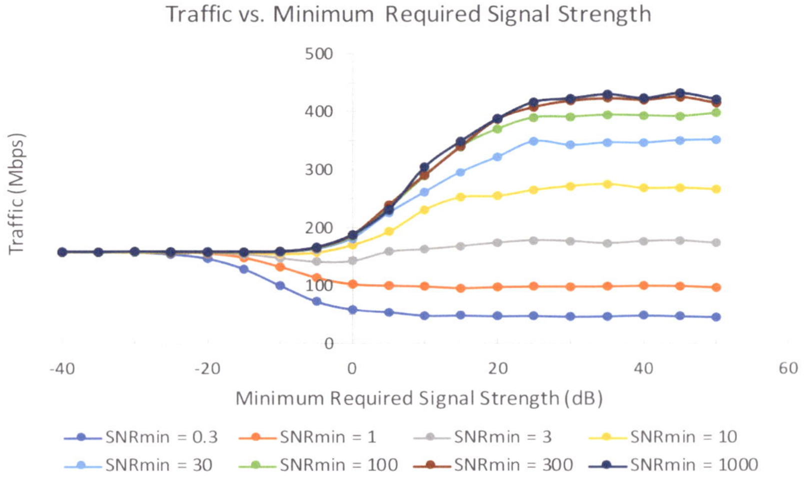

The trade-off for increasing spectral efficiency is lower coverage. In all cases, as the minimum-required signal strength (minSS) increases, fewer subscriber locations are selected to receive data. (See Figure 55 and Figure 56.) Minimum-required signal strength in dBs is defined as the sum of the minimum-required SNR (SNR_{min}) and η. (See Equation (58).) So even when the minimum-required signal strength is defined, there is a range of SNR_{min} and η that could be used, and its effects on KPIs analyzed. This was illustrated in Chapter 3 when I analyzed five scenarios where minSS = 10, and SNR_{min} and η were set to different values. (See Table 23.)

As a review, I modified the classic definition of the SINR equation (see (36)) such that all terms that aren't unique to a particular RF link are collected in a term designated as η. (See equation (47). The term that is unique to a particular RF link will have the following form:

$$10^{\frac{-\left(10\gamma \log\left(\frac{d_i}{1000}\right)+F_i\right)}{10}} \tag{71}$$

Unlike Chapter 3, the term associated with the antenna height ($-13.82\log h_b$) is incorporated into η because it is assumed that all base-station antenna heights are identical. When you plug in numbers for each of the variables in the above equation, the resulting number will not represent any physical quantity. The same argument can be made about η. Note that in the exponent of Equation (71) is the component of the median path loss from Equation (33) related to distance 'd' (i.e. $10\gamma\log(d_i/1000)$). The constant component C_0 of Equation (33) is incorporated into η. Although numbers derived from Equation (71) and η do not represent any physical quantity, they still allow us to calculate all of the important signal-to-noise ratios required for our analysis (*C/I* in (41), SINR in (47) and SNR in (51)). The reason for using η is given in section 3.4.5.

When I talk about whether a subscriber location meets the minimum-required signal strength, I am determining if the quantity derived by applying Equation (71) for that location is bigger than η by an amount specified by the minimum-required SNR. (See section 3.3 or section 5.3.) Although the number from applying Equation (71) for a particular RF link does not have a unit of measure such as mW or dBm, it does represent the strength of the signal with respect to η and comparable quantities associated with other RF links. For the purposes of this section, it would be informative to analyze the quantities derived from Equation (71), which I will refer to as signal strengths. Figure 64 shows the relative frequency of signal strengths for each subscriber location.

The smallest signal strength is greater than –20 dB. If the minimum-required signal strength is –20 dB or less, then no locations will be excluded from the simulation and coverage will be 100%.

Signal strength is not a measurable quantity, but signal-to-thermal-noise ratio (SNR) as defined as the ratio of the signal strength and η is measurable. To convert from signal strength to SNR, the value of η must be established or determined. For example, if η is set to 1×10^{-5} or -50 dB, then SNR = (signal strength) ÷ 1×10^{-5} or SNR (dB) = signal strength (dB) – (-50). The figure below shows the relative frequency of SNR for the data in Figure 64.

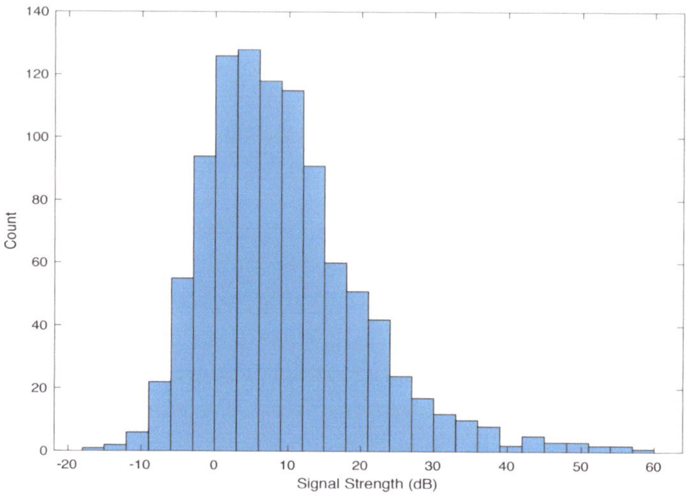

Figure 64: *Histogram of the signal strengths of subscriber locations.*

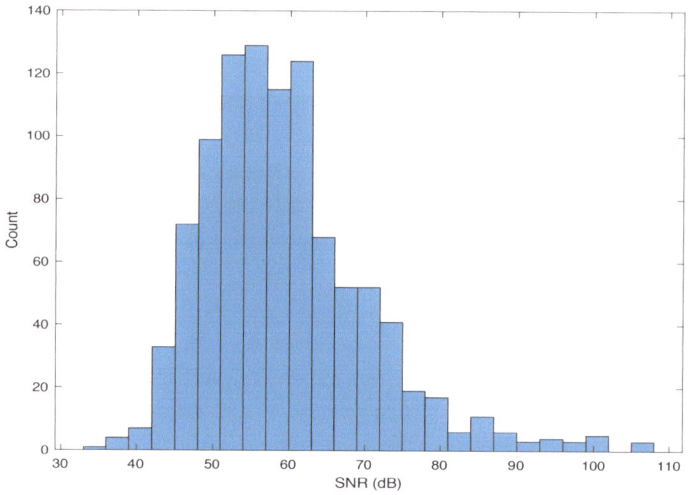

Figure 65: *Histogram of subscriber location SNRs of for η = -50dB.*

The average signal strength of Figure 64 is 9.386 dB, while the average SNR is 9.386 – (–50) = 59.386 dB, which will match the average SNR if one used the data in Figure 65 to perform the calculation. While it may appear that the average SNR can be simply calculated by subtracting η from the average signal strength (i.e. SNR_{avg} (dB) = 9.386 - η), when η is large enough, the signal strength of some of the subscriber locations may not exceed the minimum-required signal strength (minSS) and will be excluded from the simulation. For example, if η is set to 1 (0 dB) and the minimum-required SNR is also 1 (0 dB), then the minimum-required signal strength is 1 (0 dB). Examining Figure 64, all the subscriber locations where the signal strength is less than 0 will be excluded. (See below.)

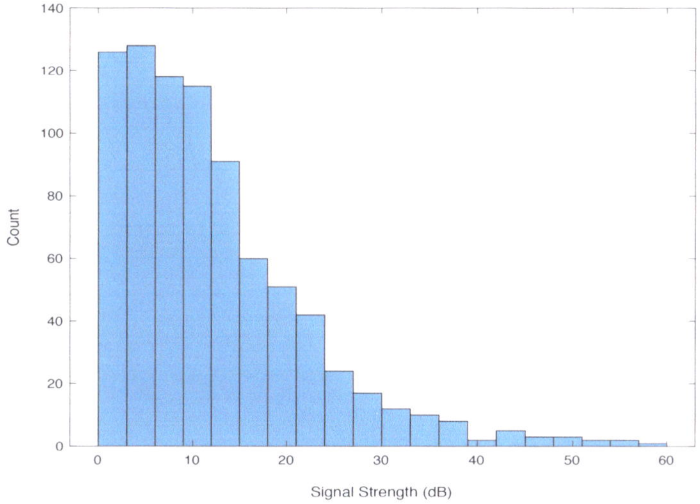

Figure 66:
Histogram of the signal strengths of subscriber locations. Minimum-required signal strength = 0 dB.

In this scenario, since $\eta = 1$ (0 dB), SNR is equal to signal strength. Hence, the histogram of Figure 66 also represents the relative frequency of SNRs. The average signal strength and the average SNR are both 12.2 dB.

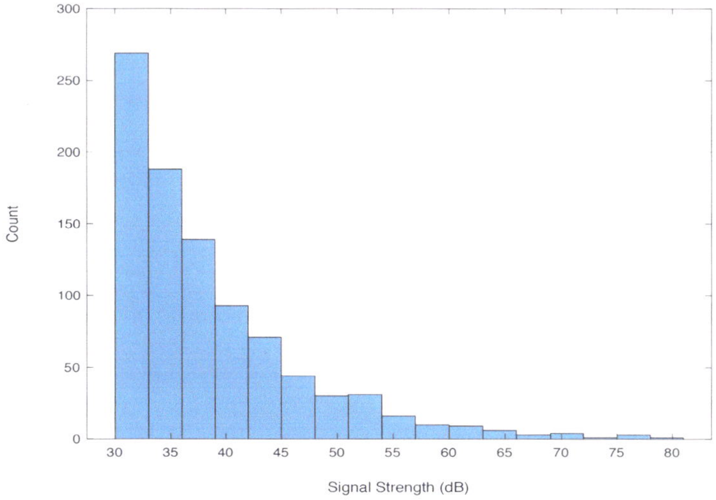

Figure 67:
Histogram of the signal strengths of subscriber locations. Minimum-required signal strength = 30 dB.

As the minimum-required signal strength increases, more subscriber locations get excluded. The histogram in Figure 67 shows the relative frequency of signal strengths when the minimum-required signal strength is set to 1,000 or 30 dB.

The average signal strength is 38.53 dB. Since minimum-required SNR is 0 dB, η is set to 30 dB and the average SNR is 8.53 dB.

For illustrative purposes, the figure below shows the subscriber locations that meet the minimum-required signal strength of 1,000 or 30 dB. Notice that the subscriber locations congregate around the cell sites. Less than 5% of the potential subscriber locations meet the minSS requirement.

Figure 68: *Cell sites locations and subscriber locations meeting minSS = 1,000 or 30 dB.*

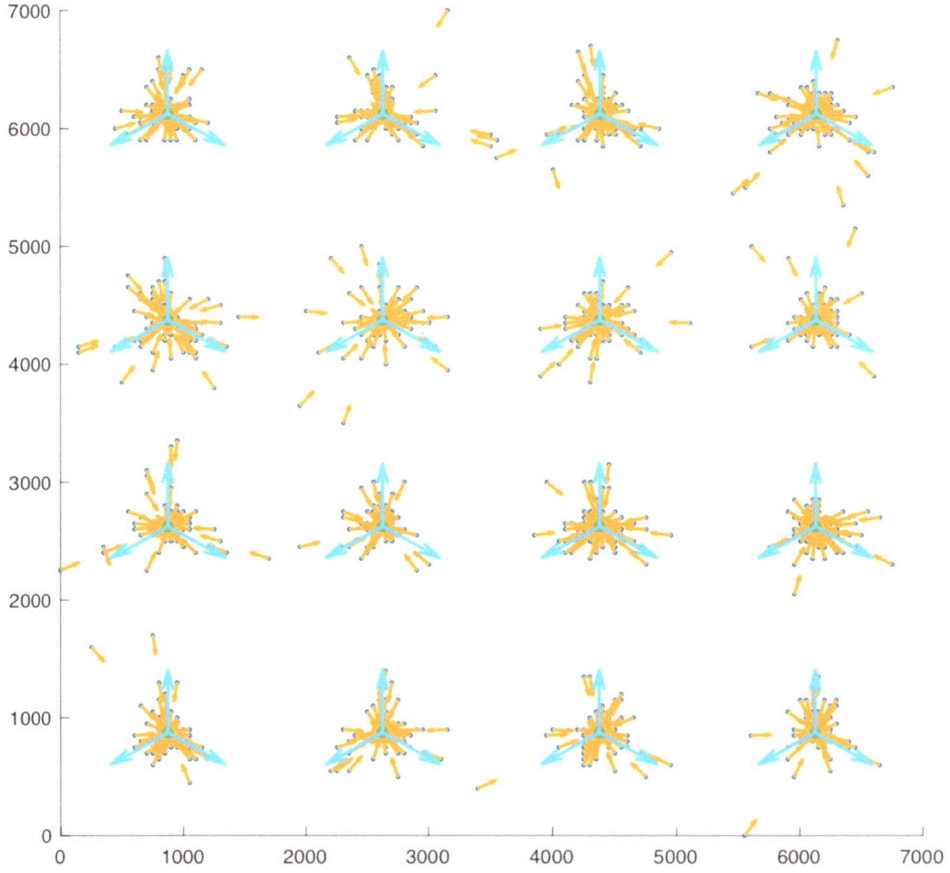

Adjusting minSS is another method of affecting the geographical distribution of the subscriber locations. Recall that when randomly placing a limited number of subscriber locations it is likely that some sectors will have more subscriber locations than others, thus impacting network KPIs. Eliminating subscriber locations based on signal strength will also affect network KPIs. The next chapter will further explore the idea of how geographical traffic distribution affects network KPIs.

The graph below shows how SNR varies with η when the minimum-required SNR is 1 (0 dB).

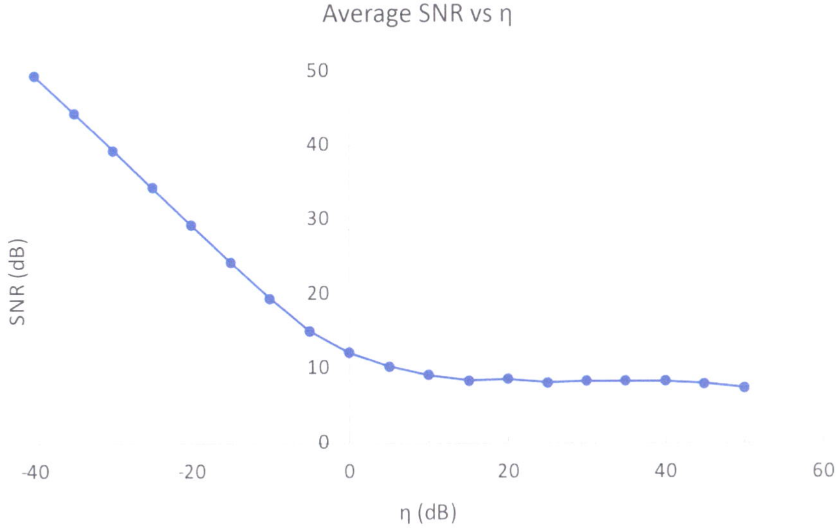

Figure 69: *Average SNR vs η for* SNR_{min} *= 1 (0 dB). When* SNR_{min} *= 1, the minimum-required signal strength equals η.*

The figure is interesting, but what we are really interested in is how SNR varies with the minimum-required signal strength. In the above figure, the minimum-required SNR is set to 1 (0dB). Thus, by using either (57) or (58), the minimum-required signal strength is equal to η and the above graph can be relabeled as SNR vs. minSS. The question is how does this graph change when the minimum-required SNR is assigned a different value? For example, let's say the SNR_{min} is assigned the value 10 dB. When minSS is equal to say -50 dB, then, from (58), η is equal to -60 dB. The signal strengths of the subscriber locations do not change under any circumstances. Thus, when η decreases by 10 dB, the SNR for each of the subscriber locations increases by 10 dB, as well as the average SNR. Hence, each of the data points in the figure above need to be increased by 10 dB. (See Figure 70.)

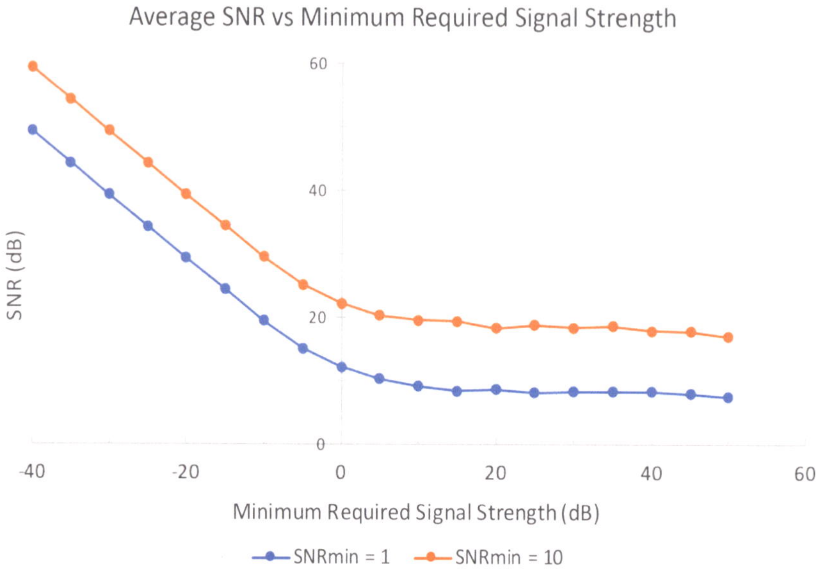

Figure 70: *Average SNR vs minimum-required signal strength for $SNR_{min} = 1$ and $SNR_{min} = 10$.*

What ultimately determines data-link speeds is SINR. SINR is a function of SNR and co-channel interference (*C/I*). *C/I* is only a function of the geometry of the subscriber locations (review section 3.2 if there is any confusion). We know from above that the subscriber locations included in the simulation will depend on the minimum-required signal strength. As the minSS increases, more subscriber locations are excluded, affecting the resulting average *C/I*. (See below.)

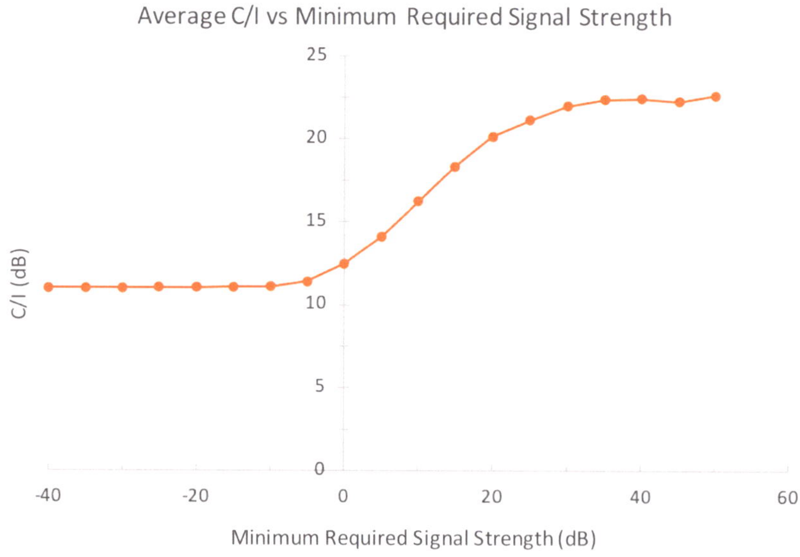

Figure 71: *Average C/I vs minimum-required signal strength.*

We now combine all the signal-to-noise ratios on one graph for a particular scenario. The first scenario is when the minimum-required SNR is set to 1 or 0 dB. (See below.)

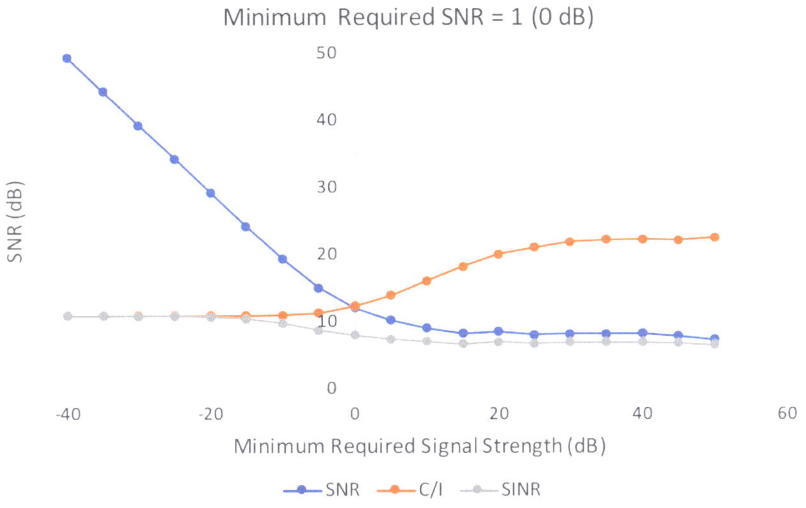

Figure 72: *Various signal-to-noise ratios for $SNR_{min} = 1$ (0 dB).*

From Figure 72 figure, it is clear the SINR curve tracks below the lower of the *C/I* curve and the SNR curve. SINR is highest for lower values of minSS, where *C/I* dominates. When SNR becomes dominant, SINR decreases, resulting in lower capacity as can be seen in Figure 63.

In the next scenario, SNR_{min} is set to 10 or 10 dB. (See below.)

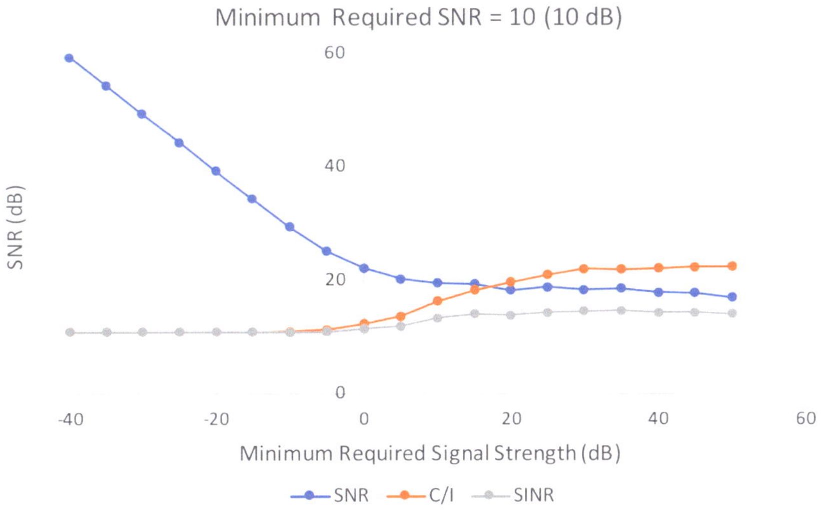

Figure 73: *Various signal-to-noise ratios for $SNR_{min} = 10$ (10 dB).*

In this scenario, SINR increases with larger minSS. The trade-off for higher SINR and data-link speeds is lower coverage.

The figure below shows the SINR curves for a wide range of minimum-required SNRs and mirrors the curves seen in Figure 63.

Figure 74: *Average SINR vs minimum-required signal strength for various minimum-required SNRs.*

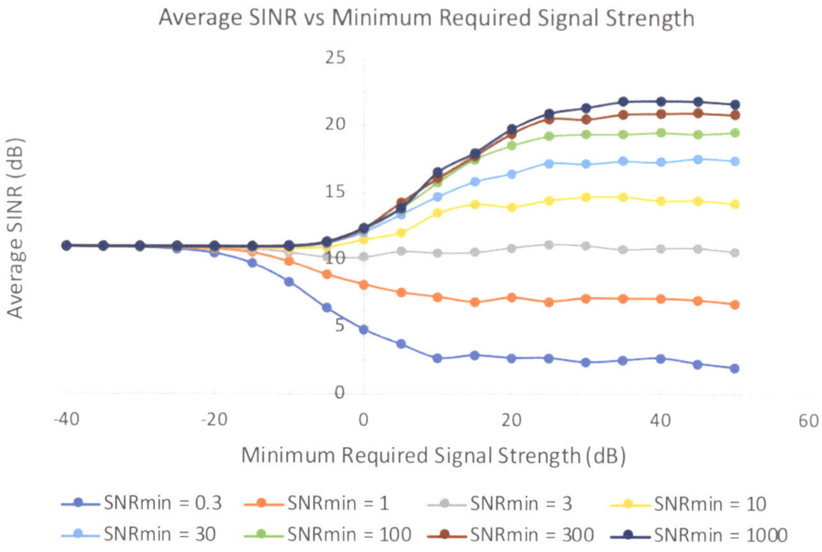

These graphs might explain why network OMs showed a higher spectral efficiency number for rural sectors than for urban. Subscribers in rural areas may only download significant amounts of data when the uplink can support higher data speeds, which means minimum-required downlink SNR is also relatively high. Rural subscribers are downloading data in a limited geographical area. In many cases, especially in rural areas, it is only desirable to cover isolated towns with one cell site. Since the population is located in a small area, it is possible to have a network that covers only a fraction of the geographical area yet serves most of the people.

One other aspect that should be mentioned is the asymptotic nature of many of the above graphs. For a given SNR_{min}, all of the above graphs show very little change in *C/I*, SNR, SINR and traffic after minSS exceeds 30 dB. After minSS exceeds 30 dB, each cell site and the subscriber locations that surround them become more isolated. (See Figure 68.) Virtually all the co-channel interference arises from the side lobes of the non-serving sectors of the cell site serving the subscriber location, and virtually none from other cell sites. If each cell site is completely isolated, average *C/I* is determined by how the subscriber locations are distributed around the cell site. The geometry of the subscriber locations is only a function of the antenna patterns and azimuths. A similar argument can be made for the average signal-to-thermal-noise ratio (SNR). The distribution of signal strengths exceeding a certain level for an isolated cell site when minSS exceeds 30 dB should resemble the histogram of Figure 67, but with the x-axis label starting at minSS. The average SNR is mainly a function of SNR_{min} (average SNR is approximately 7.5 dB higher than SNR_{min} for the antenna properties and the lognormal fading assumption used in this simulation).

5.7 Fixed Wireless Access

The analysis in the previous section that demonstrates how one could trade-off coverage for capacity seems academic for mobile wireless service, since most MNOs would consider ubiquitous coverage mandatory. However, in the case of fixed wireless access (FWA), ubiquitous coverage may not be required. Another consideration for FWA is whether to use professionally installed outdoor CPEs (customer-premises equipment) furnished with high-gain directional antennas, or indoor units with low-gain omni-directional antennas placed by subscribers, in which case the analysis in the previous section can be used. The focus of this section is on the former case although comparisons with the latter will be made.

Using a high-gain directional antenna should increase *C/I* since it is possible to increase the signal strength of the desired signal by pointing the CPE towards the cell-site sector with the strongest RF signal, while interfering signals that are not in the boresight of the antenna will be attenuated. All we need to do is calculate the capacity increase afforded by the increase in *C/I*. I will reuse much of the discussion of the previous sections, leveraging the insights gained by using the abstract constructs of minSS, SNR_{min} and η. The main difference is obviously replacing the omni-directional antenna with a directional antenna, which comprises about a dozen lines of MATLAB code.

I'll examine two scenarios. The first involves setting SNR_{min} to a low value (1 or 0 dB). In scenarios where service is coverage-limited, having such a low setting for SNR_{min} is not indicated due to its impact on capacity. However, I do not restrict myself to realistic scenarios, since running a simulation only takes a minute or two, and often I find unrealistic scenarios provide unexpected insights. The second scenario is more realistic, where SNR_{min} is set to 100 or 20 dB.

Before we get into the findings of the analysis, I want to briefly mention some housekeeping items related to the operation of the algorithm. Until now, the radio receivers at the subscriber locations used omni-directional 0 dBi gain antennas. With a directional antenna, signals that arrive in the direction of the antenna's boresight will be amplified, while all other signals will be attenuated. There are two approaches that can be used to incorporate the amount of gain provided by the directional antenna. The first approach is to increase the signal strength of the received signal by reducing the path-loss component of Equation (71).

$$10^{\frac{-\left(10\gamma\log\left(\frac{d_i}{1000}\right)-gain+F_i\right)}{10}} \quad (72)$$

The second approach is to increase the variable called Gain in the equation defining η (Equation (65) for example). In Table 35 and Table 37, the mobile antenna gain is set to 0 dB. For the FWA, the CPE antenna gain can be set to the value specified by the manufacturer. The first approach reduces the path loss for every link by 'x' dB, while the second approach reduces η by 'x' dB. In the following analysis, since all subscribers will have the same FWA CPE with 20 dB gain, the second approach is used.

Figure 75 shows how SNR varies with η when SNR_{min} is set to 1 (0 dB).

If this graph looks familiar, it should, since it is identical to Figure 69. Since the FWA CPE antenna has the same gain (0 dB) as the mobile omni-directional antenna, the signal to thermal noise ratios for both networks should be identical.

Figure 76 should be different, since it shows how average *C/I* varies with minSS (minSS is the same as η since SNR_{min} is set to 1 (0 dB)).

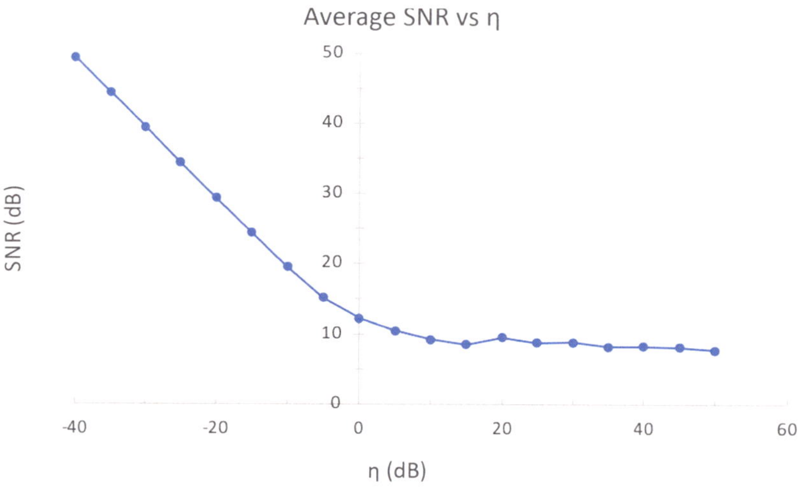

Figure 75: *Average SNR vs η for SNR_{min} = 1 (0 dB).*

The directional antenna reduces co-channel interference when coverage is at 100% (low minSS). It reduces the amount of interference from cell sites other than the one transmitting the desired signal.

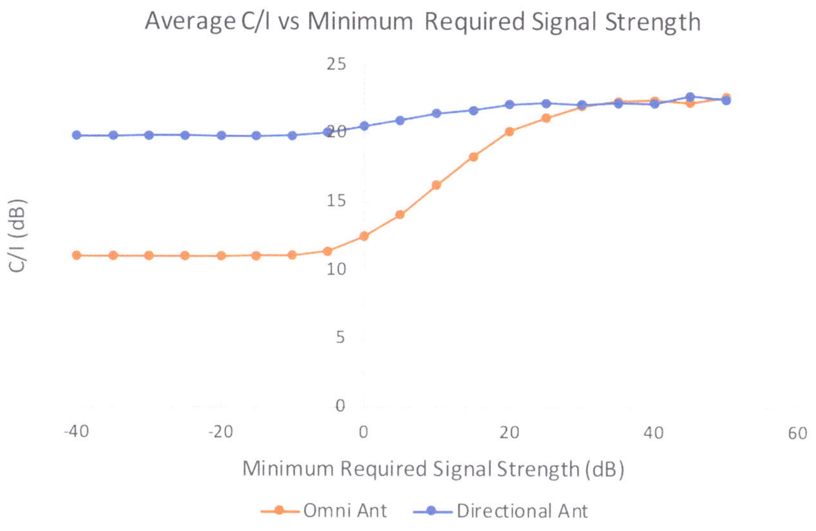

Figure 76: *Average C/I vs minimum-required signal strength.*

However, with low coverage (high minSS), the directivity of the antenna does not improve *C/I*. To see why, refer to Figure 68. Most of the subscriber locations are centered around the nearest cell site. Co-channel interference will most likely come from the side lobes of the antennas of the adjacent sectors of the cell site to which the subscriber is attached. The directional CPE antenna will not reduce interference if the interfering signal comes from the same direction as the desired signal. Notice that the *C/I* for the network using CPEs equipped with a directional antenna is nearly constant with respect to minSS.

Figure 77 shows the impact of the directional antenna on the various signal-to-noise ratios.

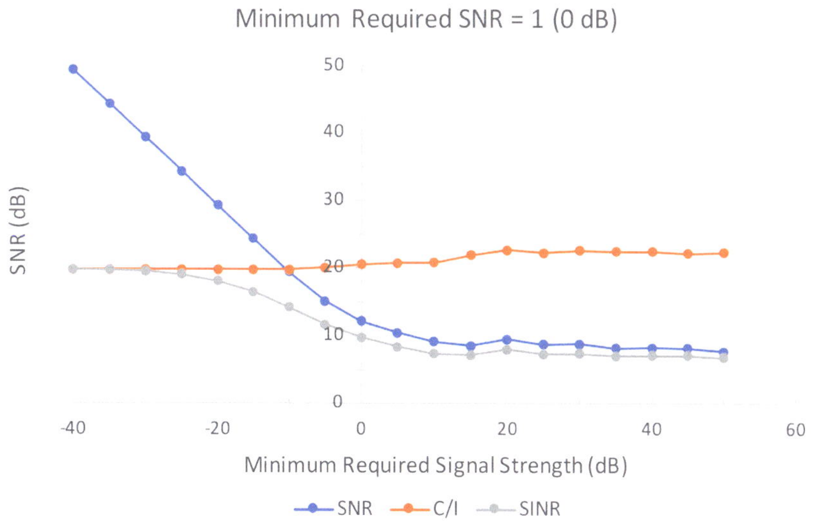

Figure 77: *Various signal-to-noise ratios for $SNR_{min} = 1$ (0 dB) using directional CPE antennas.*

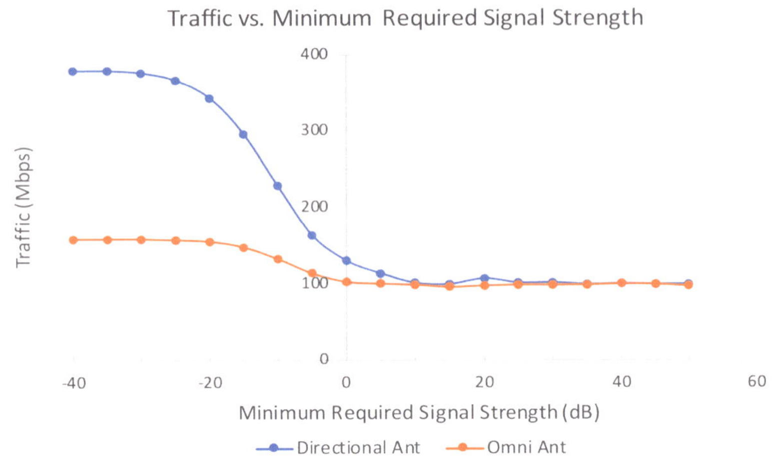

Figure 78: *Traffic vs minSS for directional and omni-directional CPE antennas. SNR_{min} = 1 (0 dB).*

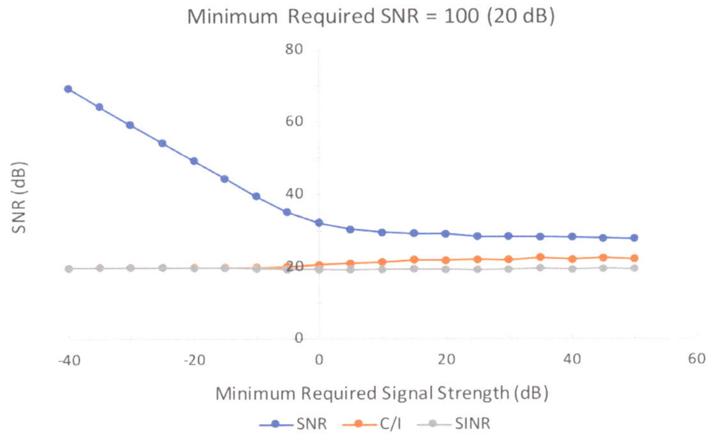

Figure 79: *Various signal-to-noise ratios for SNR_{min} = 100 (20 dB) using directional CPE antenna.*

As with the case with the omni-directional mobile antenna, co-channel interference dominates when minSS is low and thermal noise dominates when minSS is high. The improvement in traffic is shown in Figure 78.

When the minimum-required SNR is set to 1 (0 dB), there is an improvement in capacity at low minSS (urban scenario), but the directional antenna does not help increase capacity when minSS is high (remote rural areas). All three signal-to-noise ratios (SNR, *C/I* and SINR) are identical for both directional and omni-directional CPE antennas when minSS is high (compare the right sides of Figure 72 and Figure 77).

Figure 79 shows the various signal-to-noise ratios for the scenario where SNR_{min} is set to 20dB.

The SNR curve is 20 dB higher in this scenario compared to the scenario where SNR_{min} is set to 0 dB. The resulting SINR curve tracks just below the *C/I* curve, as the impact of thermal noise is not as significant as co-channel interference.

The graph below shows the traffic vs minSS for both directional and omni-directional CPE antennas.

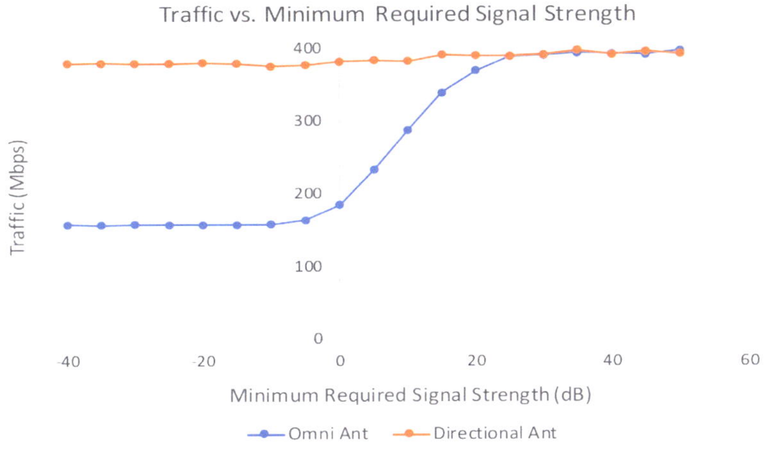

Figure 80: *Traffic vs minSS for directional and omni-directional CPE antennas. SNR_{min} = 100 (20 dB).*

As with the scenario where SNR_{min} is set to 0 dB, the benefits of directional antennas are more noticeable when minSS is low. Notice that the capacity of the network using CPEs equipped with a 20 dB directional antenna is nearly independent of minSS when SNR_{min} is set to 100 (20 dB), which is a convenient and fortuitous result.

Although the benefits of increased capacity from the use of directional CPE antennas are clear for low minSS (urban and suburban morphologies), it appears that the benefits are not as obvious for high minSS (light rural morphology). However, there are advantages from using externally mounted directional CPE antennas in low minSS conditions, even though network capacity may not be improved. The directional antenna has gain to amplify weak signals: 20 dB in our example. Also, an externally mounted antenna does not have in-building penetration losses associated with an indoor unit. Assume for example that in-building penetration losses are 16 dB. The exponent in Equation (67) could be higher by (20 + 16) / 40 = 0.9. Thus, the scale factor τ can be $10^{0.9}$ = 8 times larger for the externally mounted directional CPE antenna (the CPE can be 8 times further away from the cell site) compared to an indoor unidirectional unit. So even in cases where there are no capacity benefits of using an outdoor directional CPE antenna, there is a benefit from increased range of coverage.

Due to the relaxed physical size constraints of FWA CPEs, it is possible for them to be designed with additional receive antennas. All of the above analysis assumed the CPE can support 2x2 MIMO, and used the curve associated with 2x2 MIMO in Figure 4 to determine link speed for a given SINR. It is possible to use the 2x4 MIMO curve of Figure 4 if the CPE has four or more receive antennas. The resulting capacity increase is on the order of 50%. Some manufacturers may claim that 4x4 MIMO is possible, although field trials of FWA networks do not seem to support that claim.

One last point to consider is the use of a different path-loss exponent for wireless networks using outdoor CPEs. If line-of-sight is required for a subscriber location to be eligible for an FWA CPE, the RF path loss can be as low as 20 dB per decade .

5.8 Tool Development

I mentioned at the start of the chapter that the focus would be more on the results and less on the tool that generated the results. You may have noticed that many of the graphs have minSS (minimum-required signal strength) as the abscissa. I wanted to at least mention that I didn't run the program developed in Chapter 3 multiple times, each for a different value for minSS. It makes sense that a program designed to measure the amount of traffic at 30% utilization with differing values for minSS would perform simulations such that on average 30% of the sectors were active for each trial of the simulation. This can be done by selecting on average 14.4 subscriber locations for each trial of the simulation (30% of 48 sectors), with each simulation having a different value for minSS. The only other variable to be set before running the program is SNR_{min} (minimum-required SNR).

Looking back at Table 24 and Table 25, each column represents a different simulation, with each differing in the number of subscriber locations selected for each trial or epoch. As the number of subscriber locations change, the utilization changes proportionately until sectors become congested. For example, when the number of subscriber locations per trial is 12, utilization is 25% (12 subs / 48 sectors).

Now look at Table 39.

Again, each column represents a simulation where each differ in the minimum-required signal strength (minSS). As minSS increases, coverage decreases. As coverage decreases, the number of locations needed to be selected before being compared to minSS must increase, so that the number of subscriber locations does not fall significantly below 1,000.

5.9 Conclusion

A lot of information has been presented in this chapter, and you might feel overwhelmed. However, in many cases the analysis does not have to be complicated and some simplifying assumptions can be used in certain scenarios. Let's examine the hypothetical scenario posed in the introduction of this chapter regarding the competitor's offer of 10 MHz of low-band FDD spectrum in a dense urban area (5 MHz downlink, 5 MHz uplink). I could use the spreadsheet model in this chapter to answer this question. (See Table 38 for example.) In fact, I was playing with those spreadsheets when I stumbled upon this scenario. However, the solution can be answered without the use of those spreadsheets by utilizing some simple calculations.

Let's assume that a cell-site radio costs $20,000 and installing an antenna system for the new band is $30,000. Each sector then costs $50,000 to provision. Since a cell site has three sectors, the cost of deploying 15 MHz of downlink spectrum (3 sectors of 5 MHz) on an existing cell site is $150,000 or $10,000 for each MHz of deployed spectrum. Assume that a new cell site costs $480,000 for the shelter and base electronics, and three radios at $20,000 each are required in each sector to span the 50 MHz of the existing spectrum. The cost of installing an antenna to serve the existing 50 MHz of spectrum is the same as the cost of deploying an antenna on the offered low-band spectrum, i.e. $30,000. Thus, the total cost of deploying a new cell site is $480,000 + 3 × ($60,000 + $30,000) = $750,000 for 150 MHz of deployed spectrum (3 sectors of 50 MHz) or $5,000 for each MHz—approximately half the cost per MHz as adding the radio in the new spectrum band. The fact that the spectrum is offered in a dense urban area will negate the RF propagation advantage of the lower frequency band, thus, the spectral efficiency

minSSdB	-20	-15	-10	-5	0	5	10	15	20	25	30
trials	10000	10000	10000	10000	10000	10000	10000	10000	10000	10000	10000
spectral efficiency (bps/Hz)	0.31	0.27	0.21	0.15	0.12	0.11	0.11	0.10	0.10	0.10	0.10
spec eff active RB's (bps/Hz)	1.02	0.90	0.70	0.51	0.41	0.38	0.36	0.33	0.33	0.33	0.33
traffic (Mbps)	147.18	129.39	100.66	73.58	59.27	54.79	51.09	48.00	47.99	47.37	47.30
user perceived throughput (Mbps)	7.25	6.39	4.99	3.66	2.98	2.69	2.52	2.37	2.37	2.34	2.33
SINR (dB)	10.52	9.77	8.35	6.43	4.83	3.83	3.26	2.81	2.65	2.52	2.60
C/I (dB)	11.03	11.05	11.12	11.46	12.47	14.26	16.48	18.28	20.20	21.09	21.81
SNR (dB)	24.14	19.16	14.27	9.93	6.97	5.43	4.62	3.74	3.52	3.22	3.41
10th cell edge (Mbps)	5.14	4.65	3.59	2.51	1.85	1.88	1.85	1.84	1.56	1.78	1.72
5th cell edge (Mbps)	4.72	4.55	3.26	2.42	1.67	1.83	1.78	1.66	1.32	1.58	1.58
utilization	0.30	0.30	0.30	0.30	0.30	0.30	0.30	0.30	0.30	0.30	0.30
subscribers	1000	999	994	955	820	919	806	986	888	754	926
number of locations	1000	1000	1000	1000	1000	1500	2000	4000	6000	10000	19881
coverage	1.000	0.999	0.994	0.955	0.820	0.613	0.403	0.247	0.148	0.075	0.047

Table 39: *Network KPIs for SNR_{min} = 0.3 and utilization = 0.30.*

of the low-band spectrum being offered will be no higher than the existing high-band spectrum.

Alternatively, we can repeat the analysis from Chapter 1. With the addition of the new band, the amount of spectrum grew by 10%, thus, providing an augment in capacity equivalent to increasing the number of cell sites without the new spectrum by 10%. However, the addition of the new band increased the cost of the cell site by 20%—$900,000 with the added spectrum vs. $750,000 without. In Chapter 1, the increase in cell-site capacity could be accomplished without additional CAPEX, since it was assumed that the existing radios could be configured to operate with the supplemental spectrum as it was in the same band as the current spectrum. In this particular scenario, high-band spectrum is more valuable than low-band spectrum since the supplemental high-band spectrum does not require additional CAPEX.

Does this analysis suggest that the offer should be rejected? Not necessarily. Perhaps the government has an inventory of the same low-band spectrum that it plans to licence in an auction in the not-too-distant future. Although 10 MHz of FDD does not seem economically attractive, 20 MHz would reduce the per-MHz cost by 50%, making it equally attractive as adding a new cell site with only the existing spectrum holdings.[60] The main point of this example is to illustrate that MNOs should not automatically assume that new spectrum be acquired at any cost, even if available inventory of spectrum is exhausted. Perhaps the competitor realized that deploying this spectrum band would not be cost effective and decided to fund its own operation from the proceeds of this proposed transaction at your company's expense.

60 Alternatively, increasing the amount of spectrum by 20% would justify the 20% increased cost of the cell site.

6 | Small-Cell Analysis

Small cells are a cost-effective method to build network capacity, if done the right way. If not deployed properly, small cells can be a big waste of money and time. Small cells are the ultimate example of how the distribution of traffic effects the cost of service and key performance indicators. This chapter will also emphasize the importance of computer algorithms in formulating effective wireless strategy. I doubt there exists a manual method to effectively analyze small-cell strategies.

If one revisits the simple four-cell site network in the chapter where I introduced the stochastic method, there is no indication whether that network was a macro or small-cell network. The four cell sites could have 40 W radios utilizing high-gain 12 dBi omni-directional antennas or four small cells using 200 mW radios with 2.15 dBi quarter-wave dipoles. Both configurations can have the exact same spectral efficiency. The impact of going from 40 W radios to 200 mW radios can be counterbalanced by reducing the cell-site spacing by a certain amount. As long as the SINR of the mobile subscriber doesn't change, the spectral efficiency will be identical for both types of networks.

If one were just considering isolated small-cell networks, the material required for analysis can be found in the previous chapters. All the radios transmit the same power and all cell sectors have antennas with the same gain and pattern for the networks we have analyzed thus far. What hasn't been analyzed are heterogeneous networks with cell-site radios of various powers, using antennas with different gains and patterns. We also consider the scenario in which the mobile network operator has limited spectrum resources. This means that the operator doesn't have the luxury of dedicated spectrum for small cells and for macro cells. Perhaps it is possible to have dedicated spectrum for small cells as a temporary solution as traffic grows, but as a long-term strategy it doesn't make sense to have two different kinds of networks on two separate spectrum bands. It would be better to give the higher-capacity network all the spectrum.

In this chapter, the effects of thermal noise will be disregarded to simplify analysis.[61] In most scenarios when small cells are being considered for macro offload, the network is capacity constrained. Small cells can be used to enhance coverage inside buildings or outdoor as coverage for holes of limited size. This was the initial application for small cells and was considered the low-hanging fruit; this has since been picked. Now the more interesting application for small cells is to offload the heavily congested macro network.

61 Equivalently, η, is relatively small, thus SINR equals C/I.

It is possible that, for a particular MNO, a strategic decision was made that all new cell sites shall be small cells. This decision to prohibit the deployment of macro base stations was not accompanied by a strategy on how small cells were to be designed and deployed, a detail of importance to the cell-site planner who previously had only designed high-power macro base stations. People initially envisioned cell-site planners looking for traffic hotspots within the coverage of a macro sector and placing a small cell to offload traffic. All macro sectors must have traffic hotspots, and the cell-site planner must find them using a traffic locating tool or intuition. (Restaurants and cafes are usually clustered and must be locations where higher-than-normal data traffic is generated.) Unfortunately, when engineers started deploying small cells in this manner, they quickly discovered they did not capture much traffic. Engineers loathed designing these small cells, because it took as much time as for large macro base-station work, without providing the same benefits. For the purpose of dimensioning long-term network expansion for estimating budgets, small cells were modeled to have only 15% of the capacity of the macro sector they are designed to offload. Cell-site planners would be ecstatic if an actual small-cell capture that much traffic. Is there something inherently wrong with the small-cell model?

6.1 The Geographical Distribution of Traffic

One of the reasons given for limiting the number of randomly placed subscriber locations per cell-site sector to approximately 20 is to somewhat mimic the unbalanced traffic distribution seen in an actual network. However, it is likely that the Poisson distribution which arises when subscribers are randomly placed in a network does not capture the degree of traffic unevenness seen in a real network. There are many techniques that can create non-uniform geographical traffic distributions, such as the use of fractional Brownian motion, utilizing a population density map of a real city, estimating the location of traffic using a tool that analyzes operational measurements from cell sites, or just manually creating one. I won't go into the details of how to generate a non-uniform geographic traffic distribution, since that can be a book on its own. My philosophy is that if something can be automated, I will write a program. The MATLAB program used to produce the clustered traffic distribution in this chapter has about a dozen lines of code. The advantage of using an algorithm is that new traffic distributions can be easily generated, so analysis won't be restricted to a specific realization.

The figures below show both the network cell-site locations and the locations of the mobile subscribers for the uniform and clustered traffic distributions.

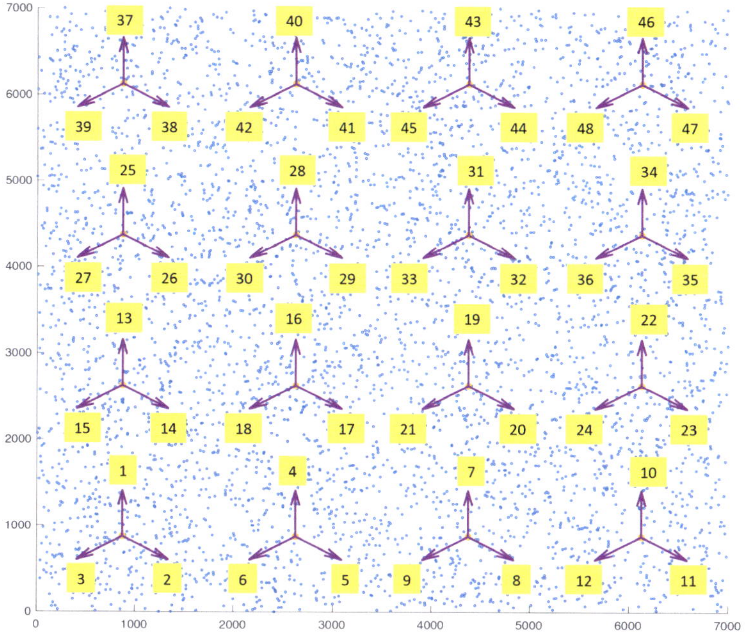

Figure 81: *Network configuration along with evenly distributed subscriber locations.*

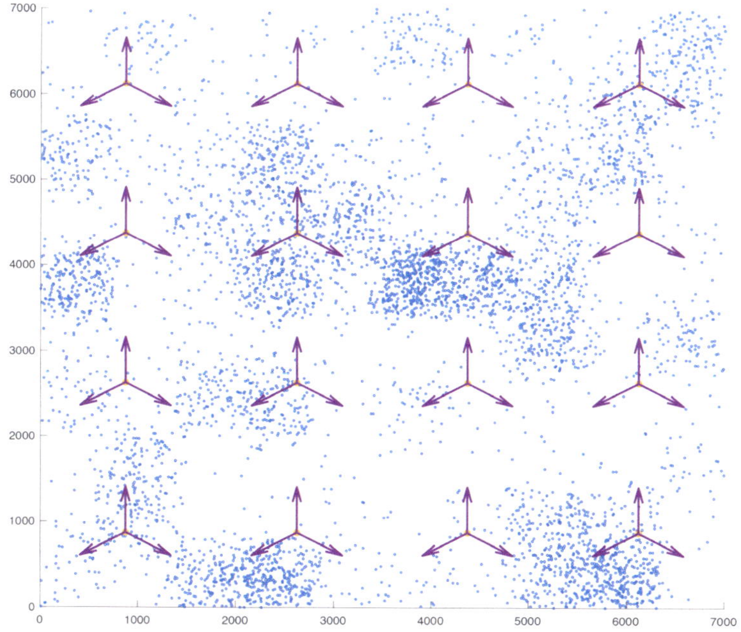

Figure 82: *Network configuration along with unevenly distributed subscriber locations.*

In the following analysis, I use a total of 5,199 subscriber locations distributed amongst 48 cell-site sectors. The reason for using so many subscriber locations per sector is that we are going to add many small cells to the network, which will reduce the average number of subscriber locations per radio from 108 to a much lower number. Also, we will be using the clustered subscriber distribution of Figure 82, which will mimic the uneven sector loading seen in a real network. The figures below shows the number of subscribers per cell-site sector.

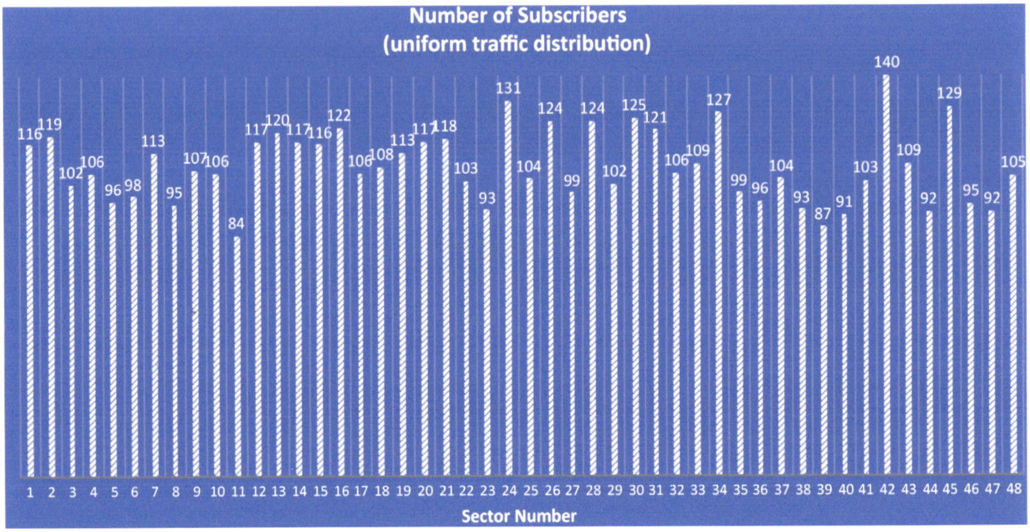

Figure 83: *The number of subscriber locations per sector for the evenly distributed scenario.*

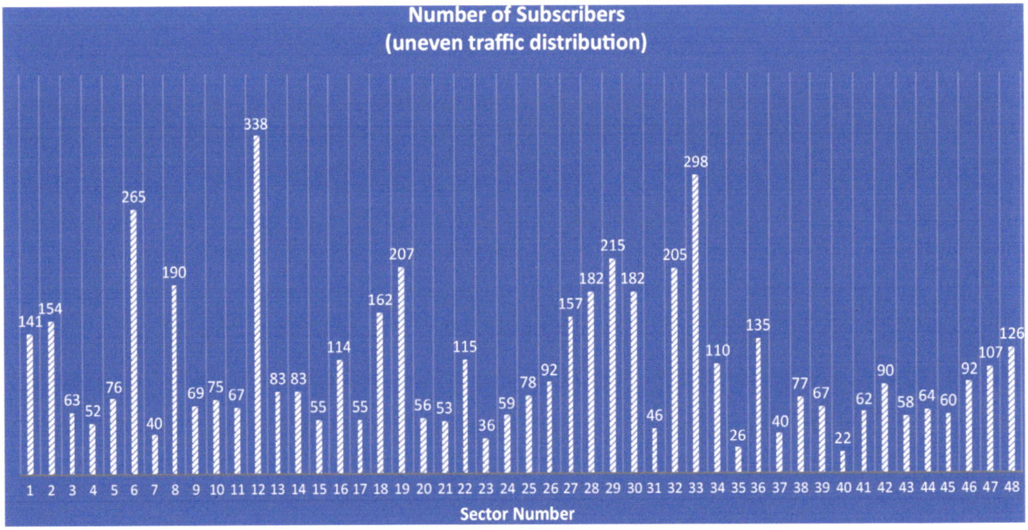

Figure 84: The number of subscriber locations per sector for the clustered scenario.

The variation in the number of subscriber locations per sector for the network with the uneven distribution is far greater than for the network with the uniform distribution. Note how Sectors 6, 12 and 33, which are pointing in the direction of the cluster of subscribers in Figure 82, have the most subscribers.

The histograms below illustrate the proportion of sectors with a specified range of subscriber locations for both clustered and uniform traffic distribution.

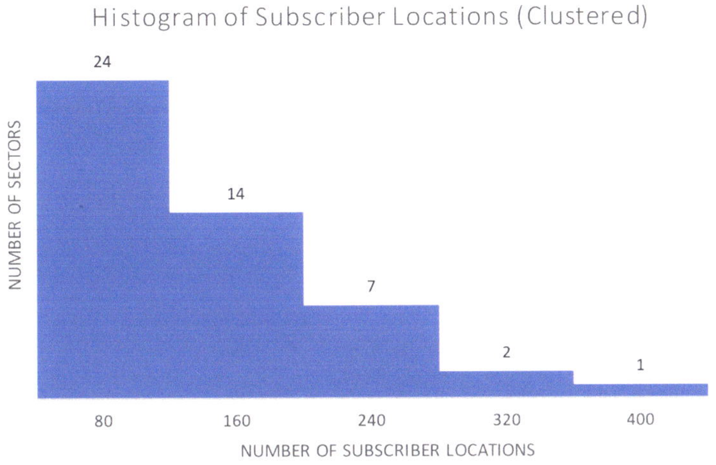

Figure 85: *Histogram of sectors with a specified range of subscriber locations.*

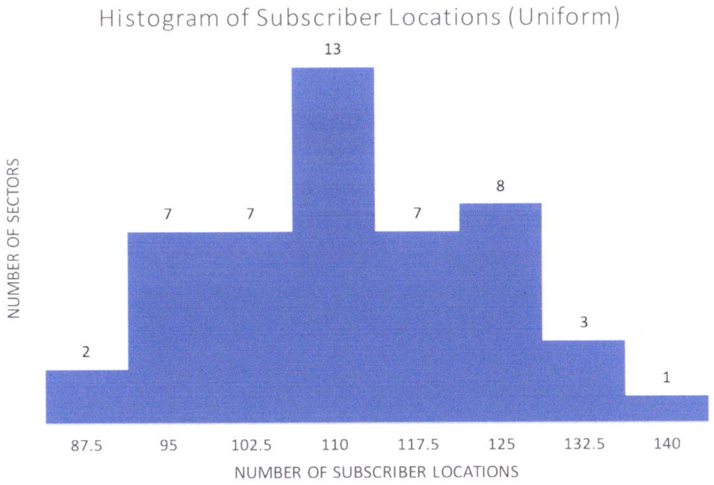

Figure 86: *Histogram of sectors with a specified range of subscriber locations.*

Note how the histogram of the uniform traffic distribution would completely fit inside the second category of the clustered traffic distribution histogram. Figure 85 resembles the histogram which shows the proportion of sectors with a given range of busy-hour data traffic or slot utilization for an actual network. The figure below shows the proportion of sectors with a given range of slot utilization for an actual CDMA network.

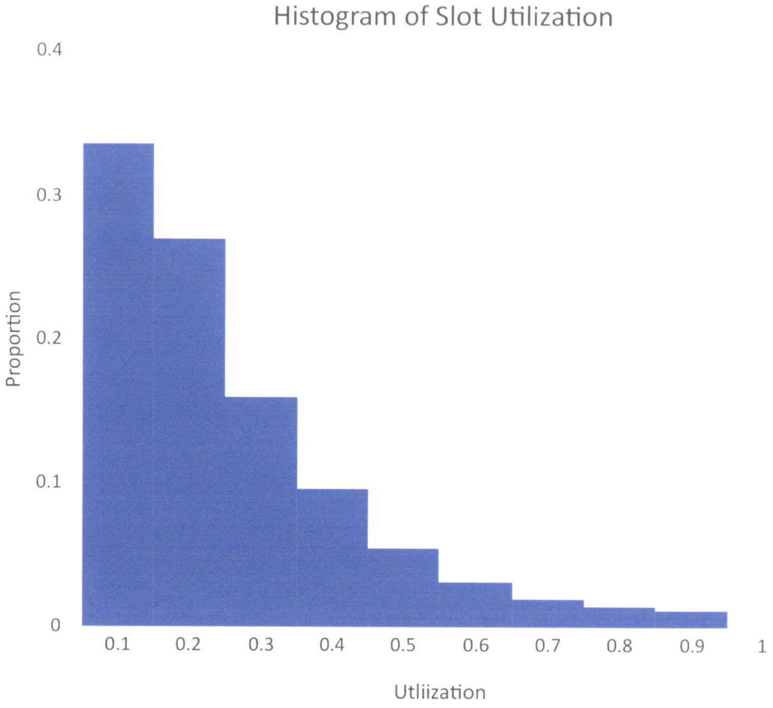

Figure 87: *Per-sector slot utilization during peak hour (mean utilization is 26%).*[62]

Although I expressed reservations about the ability of EDA to answer questions that are important in development of wireless strategy, I do believe that statistics can be used to verify models used in network dimensioning and strategy development. If the statistics of the data generated from a proposed network model match that from an actual network, one cannot exclude the application of that model. If the statistics do not match, the proposed model must be modified to resolve the discrepancies.

The effects of non-uniform traffic distribution can be seen in Figures 88 and 89.[63]

The difference in capacity when both configurations are at 100%

62 From "On Spatial Load Balancing in wide-area wireless networks," in *2012 IEEE Wireless Communications and Networking Conference (WCNC).*
63 All graphs and analyses in this chapter assume 10 MHz of LTE and 2x2 MIMO.

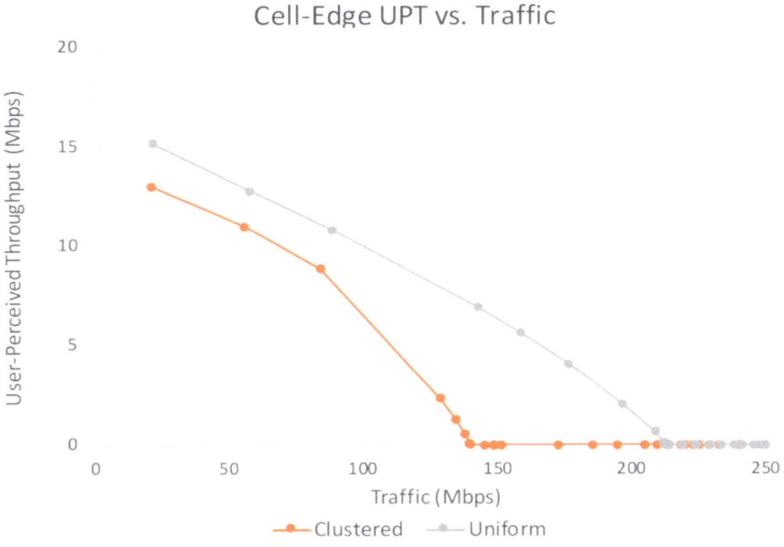

Figure 88: *Cell-edge user-perceived throughput vs traffic for clustered vs uniform traffic distribution.*

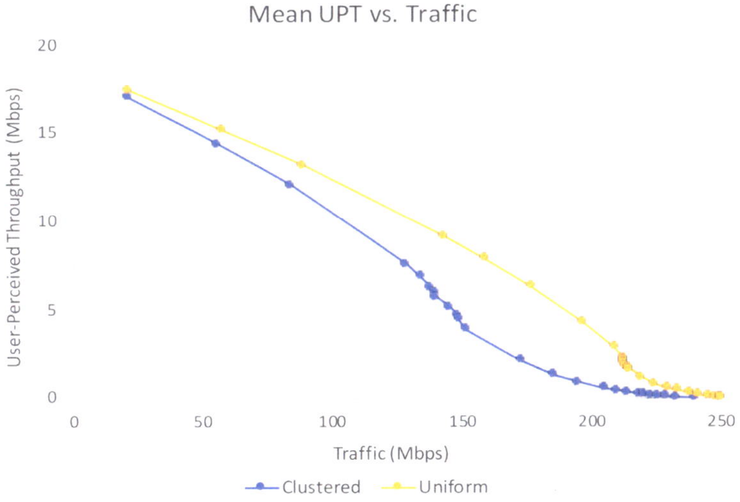

Figure 89: *Mean user-perceived throughput vs traffic for clustered vs uniform traffic distribution.*

utilization is not all that great, and fall within the range expected of random experiments[64]:

- Clustered traffic capacity: 239.9 Mbps
- Uniform traffic capacity: 249.9 Mbps

However, all the other key performance indicators at lower Resource Block utilization are worse, especially the 5th percentile cell-edge UPT.

64 Recall from Chapter 3 that traffic for a particular sector when network utilization in at 100% is uncorrelated with the number of subscriber locations served by that sector. When traffic is clustered, some sectors will carry much more traffic than others when network utilization is low (<30%). This will not affect how much traffic these sectors will carry at 100% network utilization.

Utilization	12.5%		25%		30%	
Scenario	clustered	uniform	clustered	uniform	clustered	uniform
Coverage	100%	100%	100%	100%	100%	100%
Spectral efficiency (bps/Hz)	0.176	0.186	0.269	0.298	0.287	0.331
Spectral efficiency active RBs (bps/Hz)	1.409	1.482	1.078	1.195	0.968	1.1
Traffic (Mbps)	84.6	89.07	128.9	142.8	137.9	158.7
User-perceived throughput (Mbps)	12.06	13.18	7.608	9.205	6.327	7.929
SINR (dB)	12.8	13.41	10.76	11.61	10.07	10.97
C/I (dB)	12.8	13.41	10.76	11.61	10.07	10.97
5th percentile cell-edge user-perceived throughput (Mbps)	8.87	10.84	2.343	6.944	0.535	5.658

Table 40: *KPIs at 12.5%, 25% and 30% Resource Block utilization for uniform and clustered traffic distribution.*

Since the purpose of small cells is to offload traffic from congested macro sectors, I will make a new definition for the capacity of the network. The capacity of the network is the amount of network traffic that will make the 5th percentile cell-edge UPT equal to 1 Mbps. The 5th percentile cell-edge user typically resides in the busiest sectors of the network. Hence, when the 5th percentile cell-edge UPT goes to 1 Mbps, the busiest sectors in the network will have reached nearly 100% utilization. It makes sense that the added small cells offload the busiest sectors of the network so that additional traffic can be accommodated. The capacity of the network with the clustered subscriber distribution is approximately 135.8 Mbps, while the network with the uniform subscriber distribution is approximately 205.8 Mbps. (See Figure 88.)

η is set to a very low number as I'm only interested in analyzing small-cell deployment in urban areas for the purpose of traffic offload. The definition of signal strength needs to be modified to include the gain of the base-station antenna, the height of the base-station antenna and an adjustment for the base-station power, since they are no longer the same for every RF link.

$$10^{\frac{-\left(10\gamma \log\left(\frac{d_i}{1000}\right)-13.82\log h_{b_i}+F_i+Transmit\ Power_i+Ant\ Gain_i\right)}{10}}. \tag{73}$$

The equation above is slightly more complicated than Equation (71). The path-loss exponent will be set to 4 regardless of the height of the base-station antenna, which assumes that the two-ray ground reflection model is being used.

Before analyzing small-cell strategies, I want to provide another explanation for why the spectral efficiencies of rural cell sites can be higher than the spectral efficiencies attributed to urban cell sites. The traffic distribution for a rural cluster is typically much more unevenly dispersed than in an urban cluster. For the networks depicted in Figure 81 and Figure 82, the maximum traffic of the busiest sector is 10.9 Mbps for the unevenly distributed traffic, while it is only 5.7 Mbps for the uniformly distributed traffic. Although the network capacity for the uneven traffic distribution is lower (135.8 Mbps vs 205.8 Mbps), the amount of traffic carried by the busiest sector is higher (10.9 Mbps vs 5.7 Mbps).

Network planners that rely on network OMs may assume that every cell-sector can support the maximum spectral efficiency of the busiest sector of that particular morphology. If the maximum spectral efficiency for a rural sector is 1.09 bps/Hz, the network planner responsible for dimensioning the network might assume that all rural sectors should be able to support that amount of traffic. Similarly, the network planner might assume that all urban cell sites should be able to support 0.57 bps/Hz. It is reasonable to assume that, for any algorithm that performs cell splitting at higher traffic levels, the average amount of traffic after several rounds of cell splitting should also be higher. Recall Figure 2 where the network capacity utilization was 37% for both the rural and urban cluster after 10 years of traffic growth. If a rural sector is assumed to support 1.09 bps/Hz at 100% utilization and the urban sector is only expected to support 0.57 bps/Hz then, for a given number of sectors, the rural network should support much more traffic than the urban network. This is opposite of the results obtained from the stochastic approach, which indicates that networks where the busiest sector has a very high spectral efficiency should have a lower average spectral efficiency. Perhaps the arguments of Chapter 5 can be included to explain the high spectral efficiency of both the busiest rural sector and the average rural sector as assumed by the heuristic method. It is possible that rural networks do not provide 100% coverage, and if the minimum-required SNR is sufficiently high, rural networks can have a higher average spectral efficiency than an urban network. (See Figure 63.)

I do not want to delve any further into the mystery of high rural spectral efficiency. The method I propose can be used to investigate possible explanations. More importantly, tracking the growth of the wireless network and associated KPIs will indicate whether the heuristic method can make valid predictions of future network growth. If the heuristic method comes up short, there is an alternative method that can be used in its place.

In the following small-cell analysis, the traffic distribution depicted in Figure 82 will be used and η will be assumed to be negligible (i.e. dense urban morphology).

6.2 Incremental Addition of Small Cells

Most people envisioned that small cells would be added amongst the existing macro sites, sharing the same spectrum band. The macro sites would carry the majority of the traffic, while the small cells would provide capacity offload.

Adding small cells is typically a manual process. There are so many variables, such as real estate constraints, backhaul constraints, work scheduling, etc., that attempting to automate the design and provisioning of small cells is a burden that most mobile network operators choose to avoid. Consequently, small cells are added piecemeal, one at a time, much like macro sites. In the initial deployment phase, the number of small cells may be similar to the number of macro sectors in a typical area. Even though small cells are typically added manually, I decided to take advantage of computer-aided techniques to add small cells. Figure 90 is an example of how 48 small cells were added with the aid of a computer program to the network that was analyzed in the previous section, consisting of 48 macro sectors. (See Figure 82.) I challenge anyone to do this task manually.

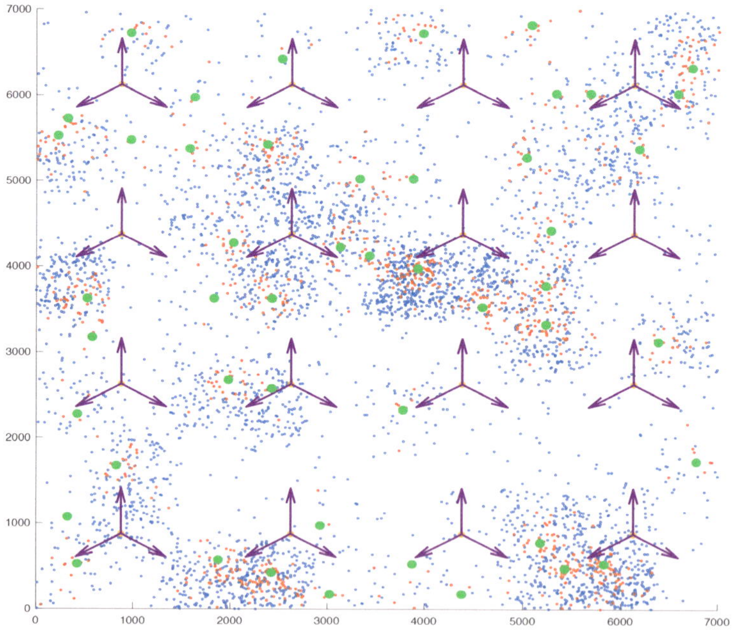

Figure 90: *48 small cells added to a network of 16 macro sites (1:1 ratio between small cells and macro sectors).*[65]

65 The small cells are represented as green dots. Blue dots are subscriber locations attached to a macro cell and red dots are subscriber locations attached to a small cell.

Since I wrote the computer program to add the clusters of traffic, it did not take a sophisticated algorithm to place the small cells. The computer algorithm added 48 clusters of subscriber locations of various sizes and placed a small cell in the middle of each. The clusters of subscriber locations were scattered amongst the 48 macro sectors, explaining the scatter appearance of small cells in Figure 90. This is the strategy advocated by industry and MNOs: find the traffic hotspot in the macro sector and place a small cell to offload traffic.

The method used to analyze small cells and heterogeneous networks is the same as in the chapter first describing the stochastic method. Since we are ignoring the impact of thermal noise, network KPIs will be solely determined by *C/I*, as described in Equation (39). In the following analysis, I'm assuming that macro base-station radios have four times the power of the small-cell radios. To simplify the analysis, I assume that the path-loss exponent γ is the same for macro cells and small cells (40 dB/decade). I include the impact of antenna heights on the path loss (the term $13.82\log(h_b)$ in Equation (45), for example). I assume that macro antenna height is 35 meters, while the small-cell antenna height is 5.6 meters. Finally, the macro base station and the small cell have different antenna patterns and antenna gains. I'm using a typical 65 degree 17.7 dBi antenna for the macro cell and an omni-directional 6 dBi gain antenna for the small cell. The resulting SINR equation looks like the following:

$$SINR = \frac{10^{-\left(10\gamma\log\left(\frac{d_i}{1000}\right)-13.82\log h_{b_i}+F_i+Transmit\ Power_i+Ant\ Gain_i\right)/10}}{\sum_{j\neq i}10^{-\left(10\gamma\log\left(\frac{d_j}{1000}\right)-13.82\log h_{b_j}+F_j+Transmit\ Power_j+Ant\ Gain_j+ABS_j\right)/10}}. \quad (74)$$

The effects on the 5th percentile cell-edge UPT can be seen in the figure below.

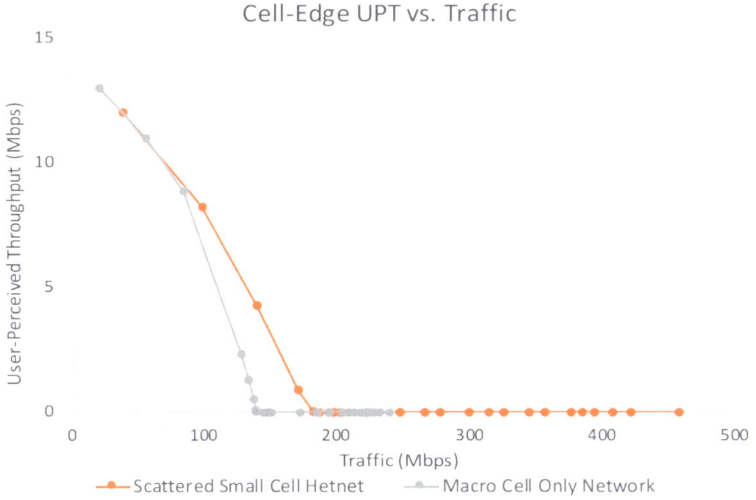

Figure 91: *5th Percentile cell-edge UPT vs traffic for both the macro-only network and the heterogeneous network.*

The network capacity has increased from 135.8 Mbps for the macro-only network to 170.6 Mbps when small cells are added (a 25.6% increase in capacity). Is this a good result? Considering that small cells are modelled to carry 15% of the traffic of a macro sector and there is a 1:1 ratio of small cells and macro sectors, a 26.5% increase in network traffic would be a fantastic result. Let's look at a few more statistics. The average number of subscriber locations per macro sector has decreased from 108.3 subscribers/sector to 88.6. The average number of subscribers per small cell is 19.7, approximately 22.2% compared to the average macro sector. Since each subscriber location downloads the same amount of traffic, the average small cell carries 22.2% of the traffic of a macro cell, which is better than the prevailing consensus of 15%. I attribute this outperformance to the fact that I know where the cluster of subscribers are located when I place the small cells—information actual cell-site planners do not have. Also note that the 26.5% increase in network traffic is greater than the 22.2% traffic offload associated with the average small cell. It is likely that small cells offload more traffic from the busier macro sectors which determine the 5th percentile cell-edge UPT.

Finally, let's analyze the KPI of the small-cell network without the macro overlay. The figure below compares the 5th percentile cell-edge UPT for three networks: macro only, macro and small-cell heterogenous network (hetnet), and small-cell only.

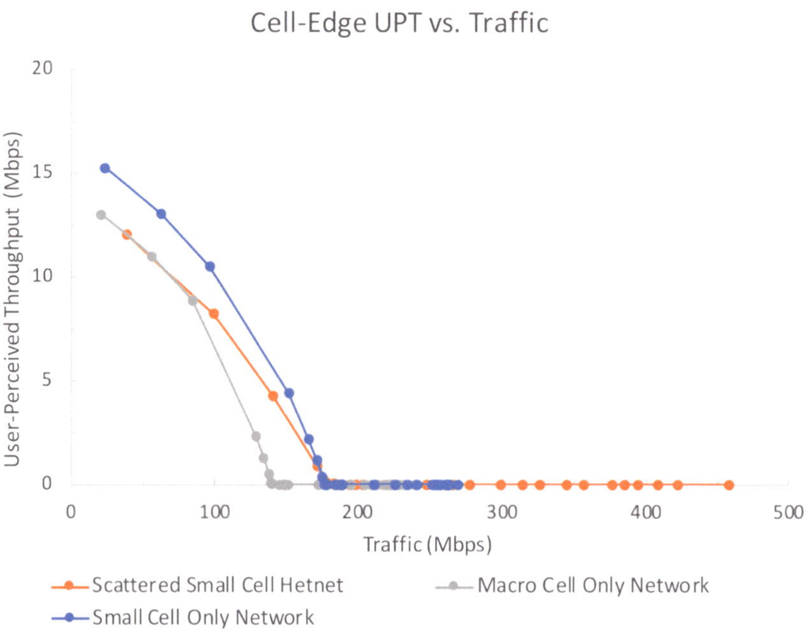

Figure 92: *5th Percentile cell-edge UPT vs traffic for macro only, heterogeneous, and small-cell-only networks.*

The small-cell-only network (capacity = 172.6 Mbps) not only outperforms the macro-only network (135.8 Mbps), it slightly outperforms the heterogeneous network (170.6 Mbps). It is probably wise to remind the reader that thermal noise was assumed to be negligible in this analysis. For this assumption to be valid, the small cells should be fairly close together. Depending on the frequency band and in-building penetration loss assumptions, the inter small-cell distance could be as low as a couple hundred meters.

There are two interpretations that can be derived from the above results:

- It is known that sectorizing an omni-directional cell site into three sectors does not provide three times the capacity, due to increased co-channel interference from the two adjacent non-serving sectors of the cell site the subscriber is using. Depending on the set of assumptions, one can derive an estimate of the reduction of the spectral efficiency due to sectorization. Based on the network configuration, antenna patterns, definition of network capacity, geographical distribution of traffic, RF path loss model and other assumptions, the above network suggests there is a 22% decrease in spectral efficiency attributed to sectorization, which can be mitigated by adding a number of small cells equal to the number of macro sectors. It is interesting to note that Andrew Viterbi quoted a 20% reduction of spectral efficiency when sectorizing an omni-directional antenna into a three-sectored cell site ($G_A \approx 2.4$ not 3).[66]
- In certain situations, the macro network does not provide any additional capacity once the small-cell network is dense enough. This strategy of removing macro sectors once there is a sufficient density of deployed small cells will be revisited several more times.

6.2.1 Optimizing Small-Cell Locations

In the previous section, it was noted that the small-cell network added 26.5% capacity to the macro-only network. Is it possible to rearrange the small-cell locations to provide greater benefits? It appears the mechanism to increase capacity is to distribute the traffic evenly amongst the macro sectors and the small cells. I wrote a MATLAB program to locate 48 small cells one at a time, to offload subscribers from the busiest macro sector. This program determines how many subscribers are attached to each macro sector, determines which sector

66 *Principles of Spread Spectrum Communication* (Andrew J. Viterbi, Addison-Wesley Publishing Company, 1995).

is the busiest, and then attempts to locate the small cell to offload as many subscribers as possible from the busiest sector. Using the clustered subscriber distribution of Figure 82, the program concentrated the location of small cells around the dense pocket of subscribers. (See figure below.)

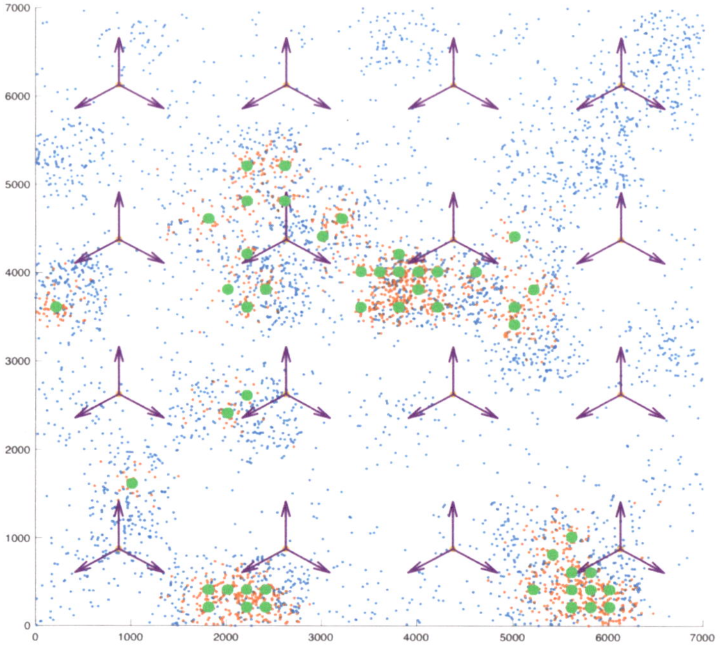

Figure 93: *Heterogeneous network (48 small cells, 16 macro sites) with more optimal small-cell placement.*

The capacity of the network is significantly higher (246 Mbps) than the network with the scattered small cell spacing (170.6 Mbps), as shown in the figure below.

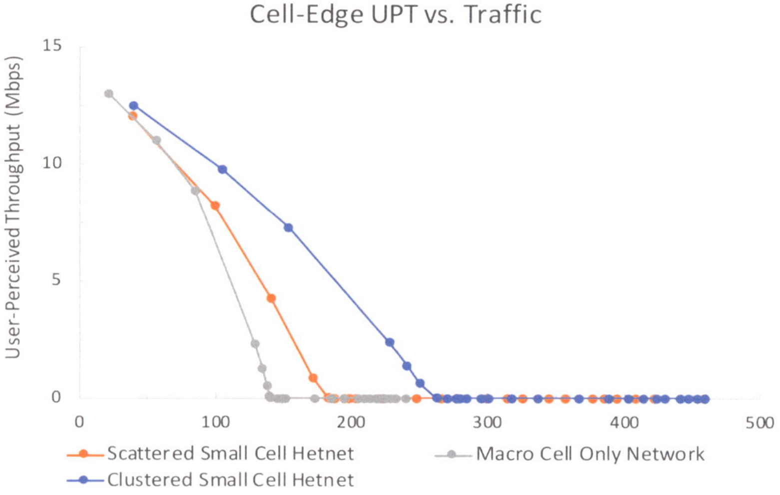

Figure 94: *5th percentile cell-edge UPT vs traffic graph comparing capacity of an optimally placed small-cell network versus a scattered small-cell network.*

As well as the clustered small-cell network performed, is there a method to make it perform even better? Unlike the scattered small-cell heterogeneous network, turning off the macro overlay is not going to improve network performance, but turning off selected macro sectors will. Knowing which sectors to turn off is not trivial, and it took investigating many different combinations to know which set of sectors should be suppressed. The problem with turning off a macro sector is that some of the traffic will migrate to another macro sector, causing it to congest sooner. The trick is to turn off all of the macro sectors that point towards a small-cell cluster; there are three such clusters in the figure below.

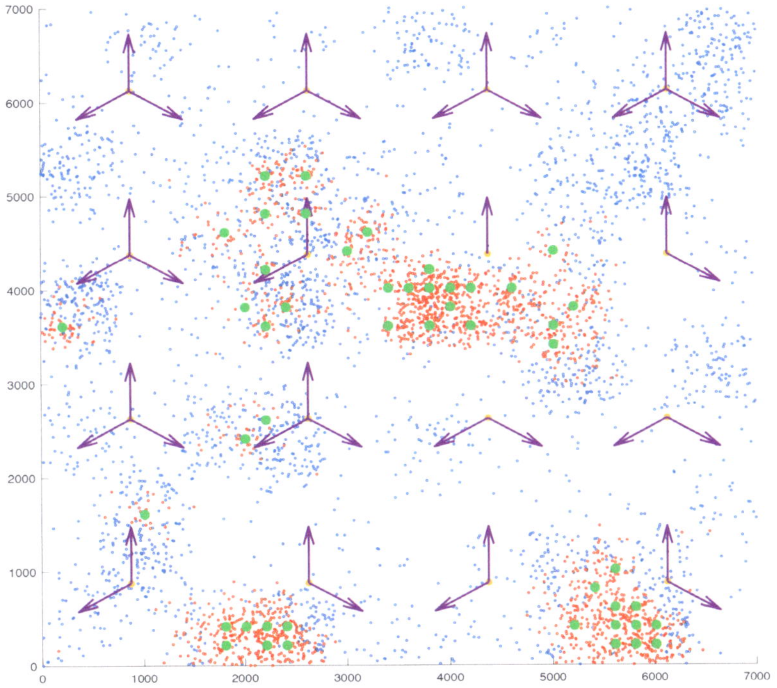

Figure 95: *Heterogeneous network (48 small cells, 16 macro sites) with more optimal small-cell placement and selected macro sectors turned off.*

The improvement in capacity is modest (259.8 Mbps for the network in Figure 95 versus 246 Mbps in Figure 93), but it does provide some insight on how to deploy small cells.

One final step to increase network capacity is to lower the power of the macro base-station radios, thus offloading more subscribers onto the small cells. (We are using the heterogeneous network displayed in Figure 93 again.) If the macro cell radios are reduced by 10 dB to 4 W, the resulting capacity of this network is 305.6 Mbps, a 125% improvement over the macro-only network. Recall that the original small-cell network only improved capacity by 25.6%. Therefore, the capacity improvement has increased five-fold by optimally placing the small cells and lowering the power of the macro cell overlay. The reason for this massive improvement is that we are converting a macro sector into a small cell. We have previously seen how the macro overlay does not increase capacity when added to a small-cell only network. However, by converting the macro cells into small cells, the small-cell only network grows from 48 small cells to 96.

It appears that the idea of scattering small cells amongst a cluster of macro cells is not such a hot idea. The solution is to locally dedicate spectrum for small cells while the remainder of the network should use macro cells. How did this idea of mixing high-power and low-power base stations in the same area come about in the first place? The Operations department is typically very suspicious of new technology, such as small cells, and can't conceive dismantling the tried-and-tested macro network. Scattering small cells amongst the dominant macro base stations seems like a safe compromise. Unfortunately, the impact of small cells as a technique to increase capacity has been a disappointment, as demonstrated in this section as well as in actual operational networks.

The choice to model small-cell capacity as 15% of the macro sector capacity as specified in the heuristic method was based on the analysis of measurements of a heterogeneous network where small cells were scattered amongst high-power macro cells. This section has demonstrated that the capacity of a small cell can be much higher—even approaching the capacity of a macro cell sector—if a more enlightened approach is used to deploy them. However, the heuristic method does not consider different values for small-cell capacity based on how they are deployed. One of the heuristic method's axioms is that all small cells have the same capacity. It is possible to modify the method to assume different values for small-cell capacity, but the selection of that value is beyond the capabilities of the heuristic method.

6.3 Ultra-Dense Small-Cell Network

As mentioned in the previous section, the concept of sprinkling a limited number of small cells amongst a constellation of macro sites is not the world's best idea. However, it will be difficult to convince cell-site planners to turn off the macro layer since it has provided the bulk of the network capacity. Clustering small cells in areas of high traffic density was a better idea than sprinkling a limited number of small cells, but the macro layer was still required to provide coverage in areas where there are no small cells. A slight improvement could be achieved by selectively turning off some macro sectors, but the example in Figure 95 did not eliminate any macro sites. If no macro sites are eliminated, it is not worth the risk of turning off some macro sectors for the limited gain that it provides.

The macro site spacing might be too large to turn-off the macro layer if the ratio of small-cell to macro sector is only one-to-one, as shown in Figure 90, and besides there is little capacity gain from eliminating the macro layer even if the inter-site distance is not an issue. Turning off the macro layer when small cells are clustered as in Figure 93 is definitely not indicated due to very poor geometry, causing low *C/I*. There is a suggestion to reduce the radio power at the macro sites, but the existing infrastructure was designed to support large high-power radios and racks of equipment requiring a room or a shelter. If the strategy is to replace all that equipment with a 1 W access point the size of a book, it would probably make more economic sense to abandon the macro site with its high OPEX costs for a lower-cost alternative. However, we will investigate the impacts of reducing the power of the macro radios by 10 dB or, alternatively, eliminating the macro network completely.

When small cells were originally proposed, it was assumed that several small cells would be required in each macro sector. The heuristic method assumes that each small cell would on average off-load 15% of the macro sector traffic, a premise which I hope you now find wanting. Using that logic, seven small cells per macro sector would be required to double network capacity. I suggest a seven-fold capacity increase is possible if seven small cells are built per macro sector only if the macro layer is turned off. Seven small cells per macro sector are a lot of small cells, and perhaps an exclusive small-cell network could be established with far fewer of them. The question is: how many small cells per macro sector are required before considering turning off the macro layer?

The next step beyond deploying one small cell per macro sector is to deploy two small cells. Figure 96 is an example of a heterogeneous network with 96 small cells in a network consisting of 16 three-sectored macros.

The addition of 48 small cells increases the capacity of the network from 246 Mbps to 320.7 Mbps (a 74.7 Mbps increase). If the macro layer is turned off, the capacity of the small-cell only network is 379.3 Mbps, a 58.6 Mbps increase over the heterogeneous network. An 18% increase in capacity is not trivial, but some may want a larger benefit before taking the risk of decommissioning the macro layer. Alternatively, the power of the macro layer radios could be reduced

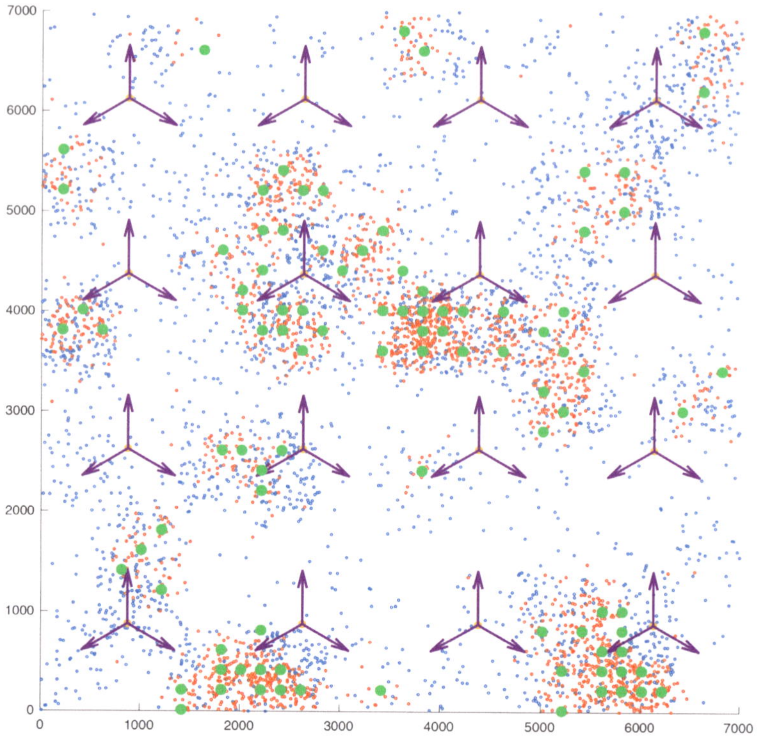

Figure 96: *Heterogeneous network consisting of 96 small cells and 16 three-sectored macros.*

by 10 dB, resulting in a network capacity of 452.2 Mbps.

I could continue this process of adding small cells, display an effective network configuration (for example in the figure above) and give the network capacity with the macro layer at full power, reduced power or no power at all, but it is more informative to show the results in the table below. Included in the table is the heuristic-model assumption that small cells have 15% of the capacity of the macro sector.

Table 41: *Capacity for various network configurations.*

Small cells per macro sector	Number of small cells	Total cells	Capacity (Mbps)			
			Heuristic -model assumption	Heterogeneous network	Small-cell only	Reduced macro power
0	0	48	135.8	135.8	0	135.8
1	48	96	156.2	246	230.2	305.6
2	96	144	176.5	320.7	379.3	452.2
3	144	192	196.9	410.7	580.5	672.4
4	192	240	217.3	491.1	723.4	736.9
5	240	288	237.7	579.1	871.9	946.7

The figure below shows the increase in capacity with the increased number of small cells.

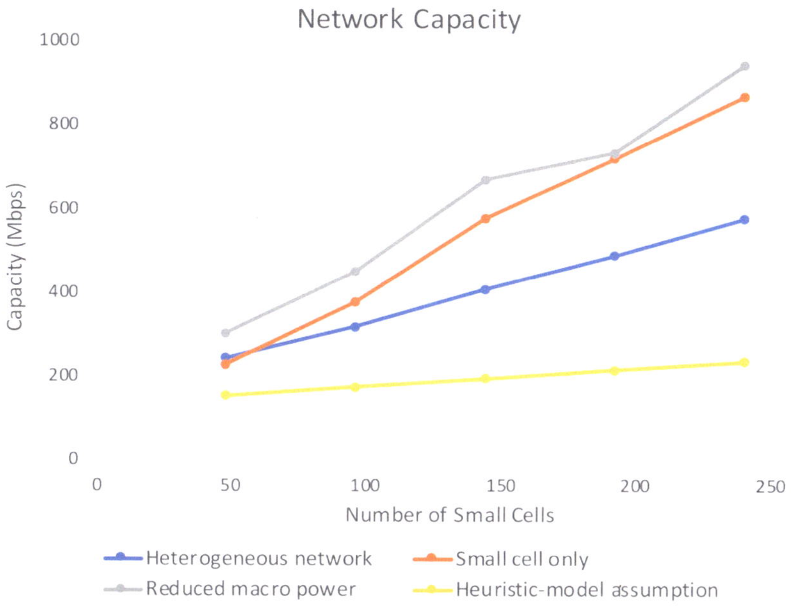

Figure 97: *Capacity for various network configurations.*

The change in the capacities for each additional 48 small cells for each of the network configurations are given in the table below.

Table 42: *Increase in capacity for each additional 48 small cells.*

Small cell number change	Increase in Capacity (Mbps)			
	Heuristic-model assumption	Heterogeneous network	Small cell only	Reduced macro power
48 → 96	20.4	74.7	149.1	146.6
96 → 144	20.4	90	201.2	220.2
144 → 192	20.4	80.4	142.9	64.5
192 → 240	20.4	88	148.5	209.8
Average	20.4	83.3	160.4	160.3

Dividing the average increase in capacity by the number of small cells added and the spectrum bandwidth, we get the spectral efficiency associated with small cells:

- Heuristic-model assumption: 20.4/48/10 = 0.042 bps/Hz
- Hetnet with small cells located effectively: 83.3/48/10 = 0.17 bps/Hz
- Small-cell only network (macro radio power = 0 W): 160.4/48/10 = 0.33 bps/Hz
- Hetnet with macro radio power reduced by 10 dB: 160.3/48/10 = 0.33 bps/Hz.

Note that the spectral efficiency of the added small cells for the small-cell-only network and the hetnet with reduced macro power is higher than the spectral efficiency of the macro-only network with the clustered traffic distribution (135.8 Mbps / 48 sectors / 10 MHz = 0.28 bps/Hz) but less than the spectral efficiency for the macro-only network with the uniform traffic distribution (205.8 Mbps / 48 sectors / 10 MHz = 0.43 bps/Hz). Since the heuristic method assumes that the spectral efficiency of a small cell is only 15% of a macro sector, I wonder how advocates of that method would react to the claim that the spectral efficiency of a small cell can be greater than a macro sector.

6.4 eICIC (enhanced Inter-Cell Interference Coordination)

Lest you think I was the first to discover the problem of scattering low-power small cells amongst high-power macro cells, the wireless telecom industry was well aware of this before the ink was dry on the first incarnation of LTE back in 2009. 3GPP's solution was to introduce a feature in LTE-Advanced called enhanced Inter-Cell Interference Coordination (eICIC). Corroborating the analysis that was just presented, 3GPP recognised small cells are unable to capture subscribers because of severe power and tower-height disadvantages compared to the overlying macro cell layer. Its solution was to make it more likely for the subscriber to select the small cell by making the subscriber choose it, even if the macro signal was stronger. This feature was called Range Extension (RE).

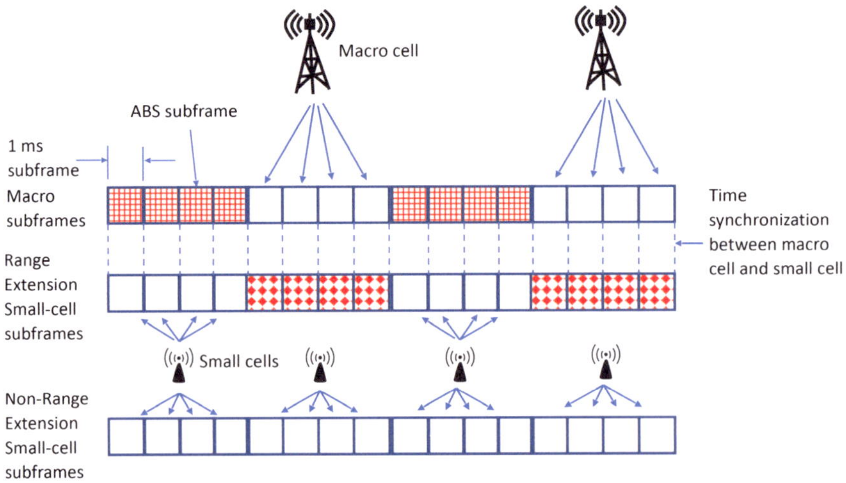

Figure 98: *The use of Almost Blank Subframes (ABS) to enable Range Extension (RE).*

The obvious problem with making the subscriber connect with the weaker small cell is that the SINR of the subscriber will be below 1 or 0 dB if the stronger macro cell is transmitting at the same time as the small cell. To solve this problem, the eICIC feature schedules the macro sector to mute its transmission on certain subframes, thus allowing small-cell users residing in the region enabled by Range Extension to experience better SINR. The muted macro subframes are called Almost Blank Subframes (ABS). The subframes are not totally blank, because the macro sector is still transmitting the cell-specific reference symbols which represents approximately 10% of the total power of the radio. The small-cell users in the area where the small-cell signal strength is higher than the macro sector do not require the Range Extension feature and can be scheduled to receive data at any time.

There are substantial implementation issues with eICIC before it can be used in an LTE network:

- All the macro sectors and small cells need to be synchronized for this feature to work. The small cell needs to know when the macro sectors are muting transmission so it can schedule the small-cell users in the region covered by Range Extension. The typical LTE network does not support base-station synchronization, and adding it is not trivial.
- Handsets support is required for eICIC to work. The mobile network operator needs most of its subscriber base to support this feature before it will have an impact on capacity.

On the surface, it seems that eICIC should be a useful feature. Why would 3GPP propose it if it weren't? Many academic research papers have been published that demonstrates the efficacy of eICIC. I have performed some analysis of this feature using the method I've been advocating and have seen some benefits. However, there hasn't been much interest in deploying eICIC among telecom operators; this is unlike Carrier Aggregation, which has seen nearly universal adoption and was proposed in the same LTE release. Instead of showing my analysis, I found 3GPP contribution R1-113842 submitted by Ericsson at 3GPP TSG-RAN WG1 #66bis. In the conclusion of its analysis of eICIC, it states:

> **Observation 2:** Superior "hetnet" system performance gains are achieved by reducing the macro output power, as indicated from the results with no ABS and 1W macros.
>
> **Observation 3:** Difficult to conclude on benefits with cell range expansion from the agreed purely interference-limited scenarios, when reducing macro output power to 1W outperforms ABS with zero power on unicast

channels, knowing that macro power reduction in real-life deployments with coverage limited users is usually not a particularly good idea.

Ericsson nearly kills both heterogeneous networking and eICIC with these observations. The only reason heterogeneous networks and eICIC have any hope is that small-cell deployments may not always be strictly co-channel interference limited. However, if the small-cell network is dense enough, users will receive adequate coverage without the help of a macro layer.

So how did eICIC garner so much attention in the mobile wireless industry? The solution was an attempt by the industry to solve a legitimate problem: i.e. how to offload more traffic from the macro cell to the small cell. However, poor offloading is just a symptom of a more important problem of how to increase the capacity of the cellular network in the most economical manner. Developers of eICIC correctly assumed that MNOs would be reluctant to reduce the capability of the macro cellular network and tried to devise a solution that had the smallest impact on the existing paradigm. Unfortunately for proponents of eICIC, cell-site design engineers who are staunch believers in the macro network would balk at the idea of giving macro subframes to the small cells, since macro base stations are already transmitting data on virtually all the available subframes and don't have many subframes to donate.

The main lesson of eICIC and small-cell deployment is that the cellular network is too difficult to analyze without proper tools. It is impossible to develop effective wireless strategy relying on thought experiments or the analysis of operational measurements without the aid of computer algorithms.

In case you are wondering if Ericsson's contribution had any impact on the development of eICIC, it had none whatsoever. If you peruse the Internet on the efficacy of eICIC, I doubt that you can find one paper or article that provides any criticism of the feature. There is little incentive to publish a negative finding on any wireless feature, especially if you work for a wireless equipment vendor. It appears that the industry created this solution first and then looked for data to support its efficacy, only to be surprised that there is a better solution: lower the power of the macro cell. Fortunately, for believers in high-power cell sites, the next chapter will analyze an antenna technology that can boost the spectral efficiency of macro cells several folds.

6.5 Small-Cell Summary

Network engineers working at the typical mobile network operator get excited when introducing a new technology that can provide capacity improvements of 10% to 15%. When talking about annual CAPEX budgets of hundreds of millions, if not billions, of dollars, 10% to 15% improvements in capacity can save the company a lot of money. How would that engineer react if there was a technique that can provide more than five times capacity improvement? This chapter showed that a proper small-cell strategy can improve the performance of the small cell by a factor of more than five times. Whereas the heuristic method assumes that the spectral efficiency of a small cell is only 15% of the macro cell, it was demonstrated that, with a more enlightened strategy, the spectral efficiency of the small cell can be higher than the macro cell.

Before we leave this section, I'll address one practical aspect with clustering so many small cells together: more frequent handoffs. I won't get into a detailed analysis of the potential problems of handing off a call from one small cell to the next. In an ideal world, where there are no latencies and signal measurements are 100% accurate, handoffs should not be an issue. In the real world, mobile network operators try to reduce the frequency of handoffs due to the increase risk of dropped calls. In the case of LTE, there are two features that can assist in reducing the rate of dropped calls due to handoffs:

- Adaptive Single Frequency Network (ASFN): ASFN is a vendor-specific feature that makes multiple base stations appear as one large base station for users located at the edge of the cell. This mimics the soft-handoff features commonly used in CDMA mobile wireless technologies.
- Coordinated Multipoint (CoMP): A family of 3GPP features that improve user throughput at the cell-edge by coordinating downlink transmission scheduling from multiple base stations.

Both solutions may slightly reduce network capacity for better cell-edge and handoff performance. Since a cluster of small cells provides much more capacity than the high-power macro layer it replaces, a small reduction in capacity is acceptable.

7 | Advanced Antenna Systems

Although SU-MIMO (Single-User MIMO) was a feature available in 3G, it wasn't until 4G and LTE that SU-MIMO was widely implemented. Likewise, although available for LTE, Advanced Antenna Systems (AAS) will only see widespread deployment in 5G networks and beyond. The classic example of an AAS is the beamformer which uses an adaptive array consisting of multiple transceiver chains and radiating elements to adaptively direct energy towards the desired target. By forming a narrow beam, the beamforming array not only increases the signal strength and SINR for the intended target, it also reduces interference for unintended recipients of the data stream. More recently, there has been much interest in a more sophisticated and spectrally efficient technology called Massive MIMO, an advanced antenna system that also uses a multi-element antenna array. Massive MIMO increases capacity by enabling simultaneous downloading of data to multiple mobile subscriber locations. Currently, there are implementation issues that prevent Massive MIMO from being deployed for 5G. This prompted the development of a simpler technology called Multi-User MIMO (MU-MIMO) which uses the adaptive array to form multiple beams thus enabling concurrent downloading of data to many users.

There are many books written on both adaptive arrays and MIMO, and it is not my intention to provide a summary of every aspect of each technology. My main focus is to provide an analysis of the capacity gains of MU-MIMO compared to simple beamforming and conventional fixed-beam antennas. I also want to demonstrate that it isn't necessary to know every detail of the design and theory of advanced antenna systems to develop strategies to optimally deploy them. This chapter will present many unpublished findings that were only discovered by modifying the computer algorithm first described in Chapter 3.

7.1 Beam Pattern Generation

Compared to the simple omni-directional quarter-wave dipole antenna, an adaptive array appears markedly more complex. Although it takes much more knowledge and expertise to design an adaptive array, the analysis to determine optimal network strategy is fortunately not that much more complicated.

In Chapter 3, where a simple four cell-site network using omni-directional antennas was analyzed, the azimuthal or horizontal beam pattern was not examined as it consisted of a simple circle. One of the drawbacks of using an omni-directional antenna is that it transmits energy in all directions in the horizontal plane, creating interference to mobile subscribers who are not the intended recipient of the data. In Chapter 5, to reduce the proportion of UEs receiving interference, directional antennas are used to transmit energy in a particular direction, resulting in increased capacity compared to a network using omni-directional antennas. The antenna pattern was obtained from a manufacturer's data sheet, but a simple formula could have been used to describe the antenna pattern. The antenna pattern can be considered a parameter which can be modified to determine its effects on network coverage and capacity. Beamforming further reduces the amount of unintended interference which should further increase *C/I* and network capacity. For the adaptive array, it is recommended that the beam pattern be derived from a formula.

For our purposes, the adaptive array is characterized by the number of horizontal elements[67] and by the selection of individual beam patterns. In the case of MU-MIMO, which allows multiple simultaneous downloads, the maximum number of simultaneous users will also be specified. The three fixed-beam antennas at each cell site in Figure 52 will be replaced by three adaptive arrays, each covering 120 degrees. It will also be assumed that the adaptive array does not have significant back lobes. In the simulation, the radiated signal strength of the adaptive array between 90 degrees and 270 degrees will be 30dB below the peak power of the main beam.

While the MATLAB command used to derive the beam pattern for an adaptive array is exceedingly simple, explaining why this command works requires a background in signal processing. For example, to generate the bore-sight beam pattern for an n-element array, the following MATLAB command can be used:

```
C = fftshift(fft(ones(1,numelements),1024))/numelements;
```

Where `numelements` is the number of elements of the adaptive array. For those who are interested in why this command works, refer to *Statistical and Adaptive Signal Processing* by Manolakis, Ingle and Kogon. As a quick aside, the main operation of the above MATLAB command is the Fast Fourier Transform (FFT) which was published in a paper in 1965 by James Cooley and John Tukey whom we have met earlier. Prior to the development of the FFT, the above command would have used the discrete Fourier transform to produce the exact same results but in a time frame that is longer by many orders of magnitude. The speed advantage of the fast Fourier transform is critical in so many applications that it

67 Some use the term "columns" instead of elements. Each column consists of a number of vertical elements to create a beam pattern in the vertical or elevation plane. In our analysis we do not consider the elevation beam pattern as we assume that all subscriber locations are located in the horizontal plane.

has been credited as among the top ten most important algorithms of the 20th century.[68] OFDM, the modulation scheme used by LTE, would not be possible without the FFT. Although, there are parts of the program where the speed of execution is important, the time to calculate the Fourier transform is not essential in my application since the operation is only repeated a limited number of times to generate the required beamformed antenna patterns.

In another coincidence, Tukey was motivated to investigate a faster implementation of the discrete Fourier transform to assist a colleague develop a beamforming array[69]. Tukey, of whose existence I was oblivious, made substantial contributions in many areas which I have unwittingly leveraged.

I used variations of the above command to generate the set of potential beam patterns for a four-element adaptive array as shown in Figure 99.

These are not the only set of beam patterns that can be formed from this array. Other than to say that they possess certain desirable properties, the reasons for the selection of these beams is beyond the scope of this discussion. Note that there are three beams to choose from if the adaptive array is to cover 120 degrees. The analysis in this chapter will assume that each beam can support two layers of SU-MIMO (i.e. 2x2 MIMO). 10 MHz of spectrum bandwidth is also assumed.

Continuing, the candidate beam patterns for an eight-element adaptive array are shown in Figure 100.

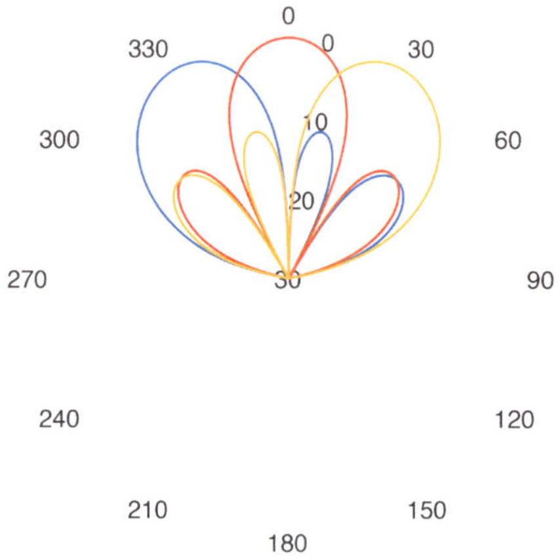

Figure 99: *Beam patterns of a four-element adaptive array.*

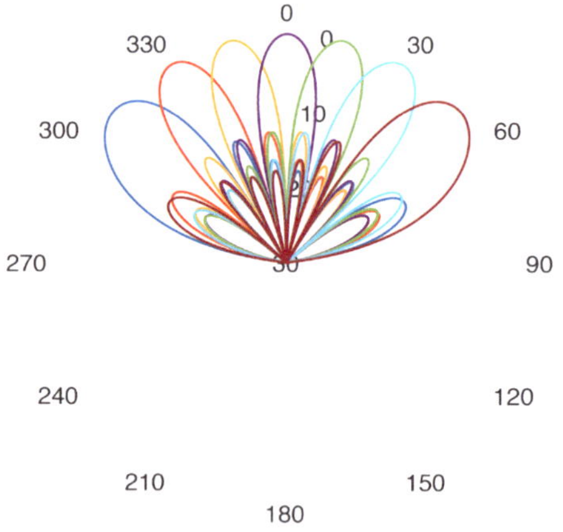

Figure 100: *Beam patterns of an eight-element adaptive array.*

68 Guest Editors Introduction to the top 10 algorithms, J. Dongarra and F. Sullivan, *Computing in Science & Engineering*, vol. 2, issue 1, pp. 22-23, January 2000.
69 James Cooley and John Tukey discuss the origin of the FFT at this url: https://ethw.org/Archives:History_of_FFT_with_Cooley_and_Tukey

Now there are seven beams that can be selected. Lastly, we will analyze an adaptive array with 16 elements. (See below.)

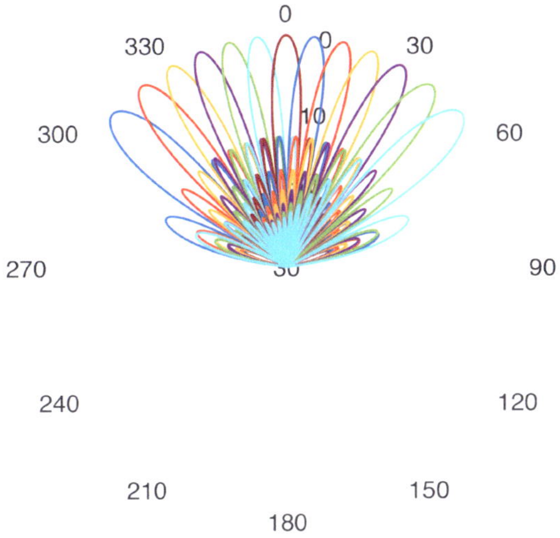

Figure 101: *Beam patterns of a 16-element adaptive array.*

There are 13 possible beams to select from.

Notice that for each of the adaptive arrays, the beams generate significant side lobes that are only 13 dB below the level of the desired beam.

The beam patterns were developed only for the horizontal plane. The analysis in this chapter assumes that subscriber locations lay on a two-dimensional plane. Although this assumption is applicable for most scenarios, for cities with a significant number of tall skyscrapers, a 3-D adaptive array that can also form beams in the elevation plane should be analyzed to determine possible capacity and coverage enhancements.

7.2 Assigning Beams to Data Buffers

In the section that describes how to derive network capacity and UPT (see Capacity, Traffic and User-Perceived Throughput – Section 2.3), we assume that there was a single buffer or queue for each sector. Although the standard beamformer can form beams in different directions, only one beam is active at a time. Hence, the standard beamformer could also be modeled as having a single input buffer or queue. (See Figure 102.) The amount of traffic and the UPT of the sector is determined by the capacity and utilization of the data buffer serving the beams.

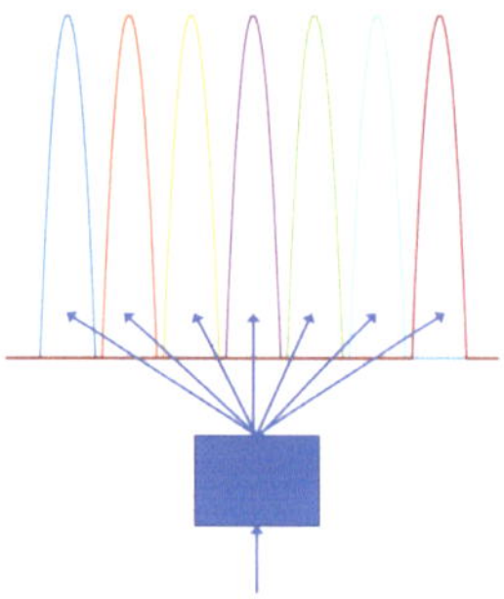

Figure 102: *Standard single queue beamforming adaptive array. At most one beam can be active at any time.*

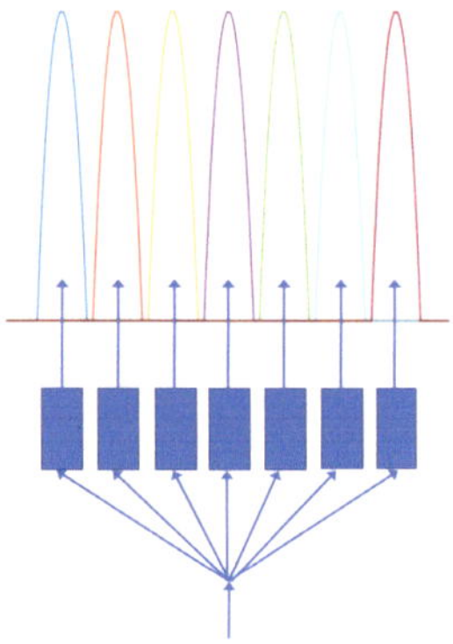

Figure 103: *MU-MIMO antenna system with as many data buffers as beams. Any beam can be active at any time.*

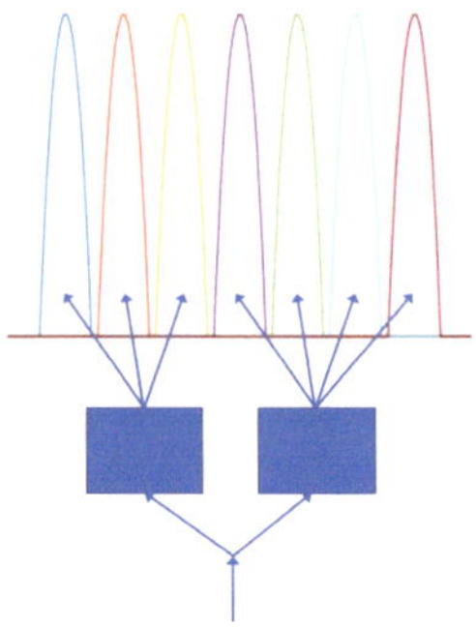

Figure 104: *MU-MIMO antenna system with two data buffers supporting at most two active beams at any time.*

MU-MIMO supports multiple beams simultaneously with the goal of increasing network capacity. The number of beams supported by the adaptive array determines the maximum number of simultaneous data streams. To calculate KPIs when the maximum allowable number of simultaneous data streams equals the number of beams, each beam should be modeled to have its own data buffer. (See Figure 103.)

There might be implementation issues supporting so many beams simultaneously, hence, it would be interesting to determine network KPIs when the maximum number of simultaneous data streams supported by the adaptive array is less than the number of beams. For example, a maximum of two active beams are supported by the adaptive array in the Figure 104.

When supporting multiple beams, the power of each beam is reduced such that the total power of the adaptive array remains unchanged. For example, when two beams are transmitting data, the power of each beam is reduced by half. This is independent of the maximum allowable number of simultaneous beams the adaptive array may support.

When there is only a single data queue (as in Figure 102) or when there are as many data buffers as there are beams (as in Figure 103), there are no ambiguities in assigning beams to a data buffer. However, when the maximum allowable number of active beams are fewer than the total number of beams, there are different strategies that can be used to assign beams to data buffers. The manner I chose to assign the beams would make the number of subscriber locations associated with each buffer as similar as possible. Also, the beams assigned to a buffer are contiguous. The biggest potential source of interference for a subscriber location are the beams adjacent to the one serving it. If all the beams in a buffer are adjacent, allowing only one beam in a buffer to be active would potentially eliminate one of the greatest sources of interference.

7.3 Impact of Beamforming on Network KPIs

Initially, advanced antennas systems will be deployed in dense urban areas as base stations in those areas will experience traffic congestion soonest. We will set the thermal noise factor η to an arbitrary low value (1×10^{-5} for example) to represent a scenario where co-channel interference is dominant, which is typical of a dense urban scenario. Since we are ignoring the impact of thermal noise, network KPIs will be solely determined by *C/I*, which is reminiscent of the analysis of small cells in the previous chapter. We will also take advantage of the macro cell site configuration and uneven subscriber locations employed in the analysis of small cells as displayed in Figure 82.

To reduce the amount of co-channel interference, the beams of 5G advanced antenna systems do not transmit cell-specific reference symbols. Hence, to isolate the effects of beamforming and MU-MIMO, the network using conventional antennas will also have radios that do not transmit the cell-specific reference symbols when data is not being transmitted. Once the KPIs of the conventional network has been obtained, the 65-degree base-station antennas will be replaced with either the beamforming antenna or the MU-MIMO antenna.

We will use the same definition of network capacity employed in the analysis of small cells, i.e. the amount of network traffic that will make the 5th percentile cell-edge UPT equal to 1 Mbps. This will facilitate a comparison between small-cell strategies and AAS deployment as both these techniques were developed to increase network capacity.

Although Figure 88 displays the cell-edge UPT for the network configuration in Figure 82, that graph assumes that cell-specific reference symbols are transmitted in every subframe regardless of whether the PRB contains data. The figure below shows the 5th percentile cell-edge UPT with and without the presence of the CRS.

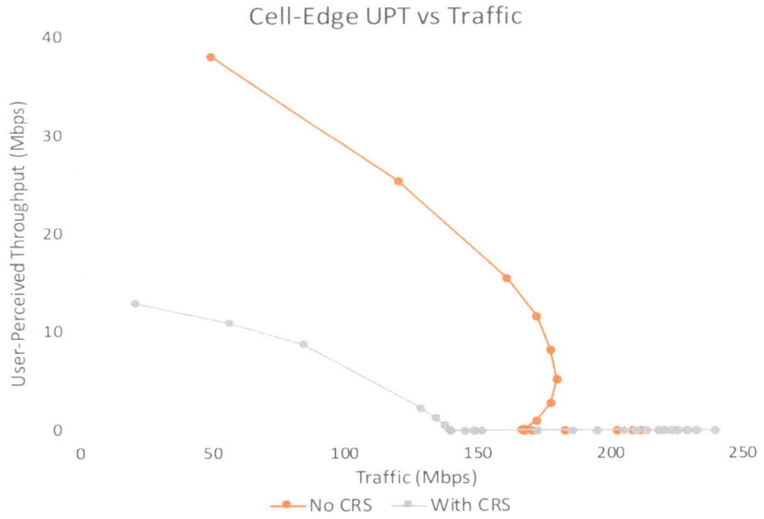

Figure 105: *Cell-edge UPT vs network traffic graph with and without CRS.*

Although difficult to see, the amount of traffic supported at 100% PRB utilization does not change when the CRS is eliminated (240 Mbps). What does change is the cell-edge UPT vs traffic graph. As seen in section 3.6, when the CRS is removed, the UPT graph develops an unusual shape. In Figure 105, when network traffic is above 166 Mbps and below 180 Mbps, there are two possible cell-edge UPTs. Fortunately, there is only one traffic level where the cell-edge UPT equals 1 Mbps which defines network capacity for the purpose of comparison: 172.1 Mbps for the network without the CRS. However, it is not recommended to operate the conventional-antenna network above 166 Mbps since there is the possibility that the cell-edge UPT will fall rapidly to 0. The strategy of reducing the level of co-channel interference by sequestering resources that are intended to carry traffic will reappear in the discussion on MU-MIMO deployment policy.

We will first analyze the performance of the network in which the conventional fixed-beam antenna is replaced by an eight-element adaptive array with the beamforming pattern illustrated in Figure 100. The figure below shows the dramatic improvement in cell-edge UPT.

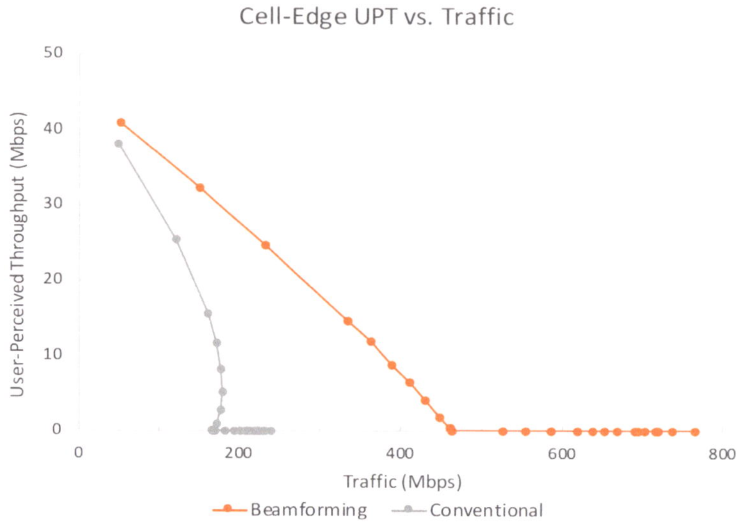

Figure 106: *Cell-edge UPT for a network using eight-element adaptive array beamforming antennas versus a network using conventional antennas.*

Not only is there a 165% improvement in the network traffic when the cell-edge UPT is 1 Mbps (455.3 Mbps versus 172.1 Mbps), but the shape of the UPT graph has implications on how much traffic the MNO may opt to support on the network. Since operating the network with conventional antennas at traffic levels above 166 Mbps is not recommended due to network stability reasons, it could be argued that a fairer comparison would be to examine the traffic levels when the cell-edge UPT is at 13.8 Mbps. The increase in network traffic with the adoption of beamforming is now 106% (342.6 Mbps versus 166 Mbps). However, the network with beamforming antennas gives the MNO greater flexibility in sacrificing cell-edge UPT for more traffic. In this example, the MNO could support 455.3 Mbps (174% increase versus 166 Mbps) at the expense of lower cell-edge UPT (1 Mbps instead of 13.8 Mbps).

For reference, the figure below shows the cell-edge UPT versus network traffic graphs for the four-element, eight-element and 16-element adaptive arrays shown in Figure 99, Figure 100 and Figure 101 respectively. Also included are graphs for a 32-element and a 64-element array.

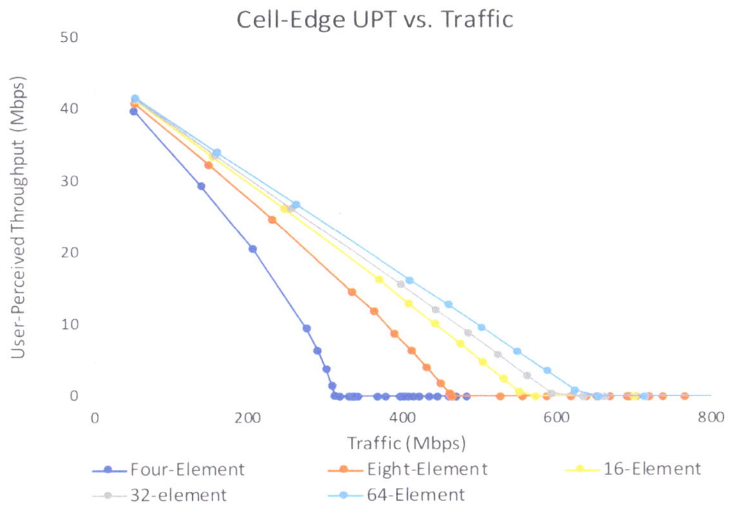

Figure 107: *Cell-edge UPT versus network traffic for various beamforming antenna configurations.*

The traffic levels at 1 Mbps cell-edge UPT are 310.2, 455.3 and 546.3 Mbps for the four-element, eight-element and 16-element adaptive arrays respectively. There are diminishing gains from adding more antenna elements as 32-element and 64-element arrays only support marginally more traffic (586 and 622.2 Mbps respectively).

Since all beamforming antennas have only one beam active at a time, the only way to increase capacity is to increase data-link speeds, which is accomplished by improving SINR. As the beams become narrower with an increasing number of antenna elements, active beams are less likely to cause interference to other subscriber locations receiving data. The figure below shows the improvement of SINR with an increasing number of antenna elements. Network capacity occurs at around 30% network utilization.

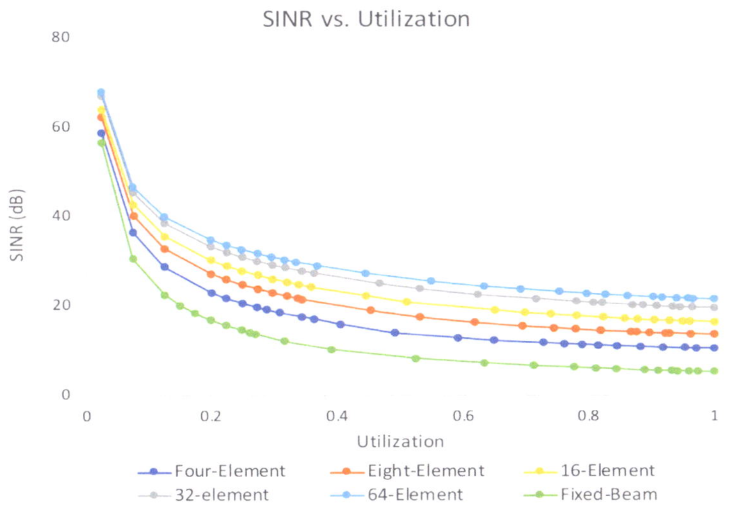

Figure 108: *SINR versus network utilization for various beamforming antenna configurations.*

In practice, 32-element and 64-element adaptive arrays will likely never be deployed. The formula for approximately calculating the beamwidth—which is typically defined as the angular extent in which the main beam is within 3dB of the maximum gain—can be found in textbooks:

$$\Delta\varphi_{3\,dB} \approx \frac{360^o}{(\pi \cdot num\_element)}. \tag{75}$$

I have a slightly different method for calculating beamwidths. The number of beams between 0 and 30 degrees is a quarter of the number of antenna elements. (See Figure 99, Figure 100, and Figure 101). Hence, there are 8 beams within 30 degrees for a 32-element adaptive array and 16 beams within 30 degrees for a 64-element adaptive array. The beamwidths for the 32-element and 64-element arrays are 3.75 degrees and 1.875 degrees respectively. Equation (75) gives slightly smaller numbers for the beamwidths but in either case the beams are very narrow. To give some context, there are 6 degrees between the ticks representing minutes on a clock face. Fortunately, the capacity gains from using a 32-element or 64-element adaptive array are minor, so little is lost by restricting the number of horizontal elements to 16 or less.

7.4 The Capacity Gains of MU-MIMO over Beamforming

The expectation is that MU-MIMO, with its ability to transmit multiple data streams over separate beams simultaneously, will have much higher capacity over both the conventional fixed-beam antenna and simple beamforming where at most one beam is transmitting data. For example, if a 16-element array MU-MIMO antenna is capable of supporting 13 simultaneous beams, what is the network capacity increase of using such an antenna over a simple beamforming antenna capable of supporting only one beam? Is it 13 times more capacity? For our analysis, we will use the same base station and subscriber locations as in the previous section. (See Figure 82.)

7.4.1 Eight-Element Adaptive Array MU-MIMO

We first examine the eight-element MU-MIMO configuration as it represents one of the most popular designs in the sub-6 GHz spectrum band. To ease into the analysis, we will examine the capacity gain from supporting at most two data streams as illustrated in Figure 104. The fact that the two-buffer MU-MIMO AAS has twice as many PRBs compared to the simple beamforming antenna should have some impact on network capacity. To view the impact, see Figure 109.

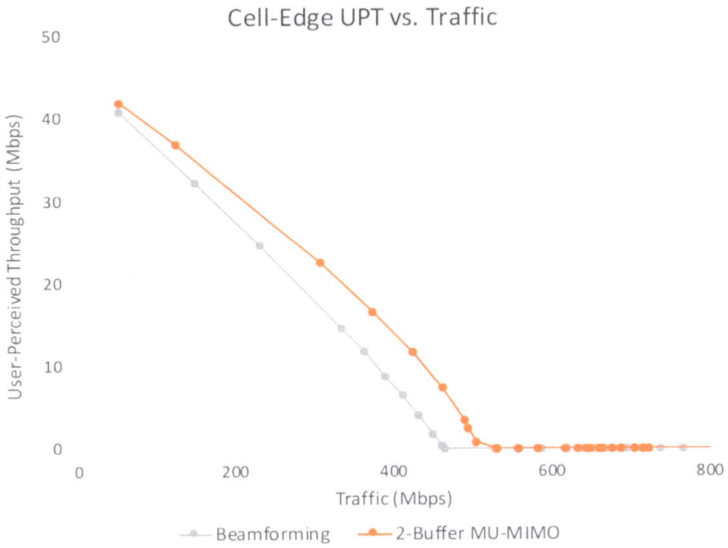

Figure 109: *Cell-edge UPT versus network traffic for two-buffer MU-MIMO and simple beamforming.*

If network capacity is defined as the amount of traffic when the cell-edge UPT is 1 Mbps, the capacity gain from adding a second data stream is 10.3% (502.3 Mbps vs. 455.3 Mbps). If the 10% capacity gain from adding a second data stream seems unimpressive, perhaps the potential of adding up to five more beams will improve the prospects of MU-MIMO. The figure below shows the network capacity gain of adding a third data stream.

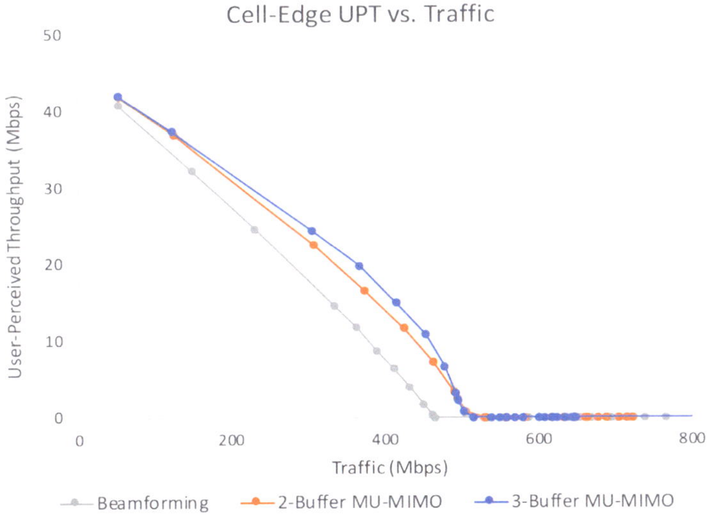

Figure 110: *Cell-edge UPT versus network traffic for eight-element MU-MIMO and simple beamforming.*

Although the cell-edge UPT is slightly improved with the addition of the third beam, the level of network traffic when the cell-edge UPT is at 1 Mbps is actually less for the three-buffer MU-MIMO configuration than for the two-buffer MU-MIMO configuration (500.3 Mbps for the three-buffer MU-MIMO).

To complete the analysis, the figure below shows the cell-edge UPT graphs for the other MU-MIMO configurations.

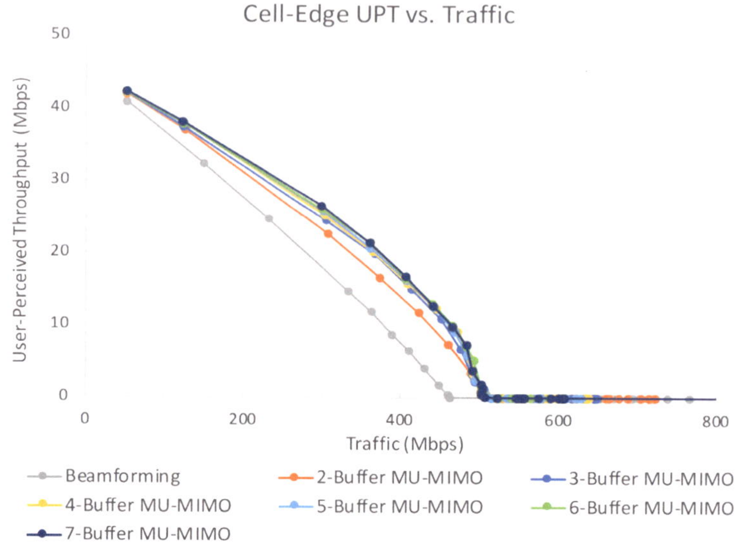

Figure 111: *Cell-edge UPT versus network traffic for eight-element MU-MIMO and simple beamforming continued.*

The best that can be said is that the additional streams do not make matters worse.

Although the addition of data streams increases the number of PRBs, the downsides of having more active beams are higher co-channel interference and lower SINR. (See below).

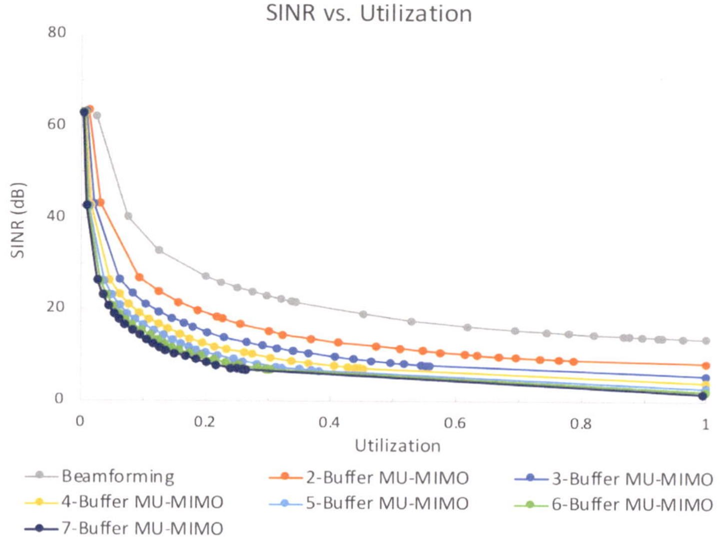

Figure 112: *SINR versus network utilization for various MU-MIMO antenna configurations.*

It is difficult to overstate the importance of selecting which KPIs are used when measuring the effectiveness of new wireless features. The KPI we chose to compare various network configurations was the maximum amount of network traffic that can support 1 Mbps 5th percentile cell-edge UPT. We could have selected another KPI such as the amount of network traffic at 100% PRB utilization. Depending on which KPI is used, one can drastically alter the appeal of increasing the number of simultaneous active data beams in a MU-MIMO antenna system as shown in the figure below.

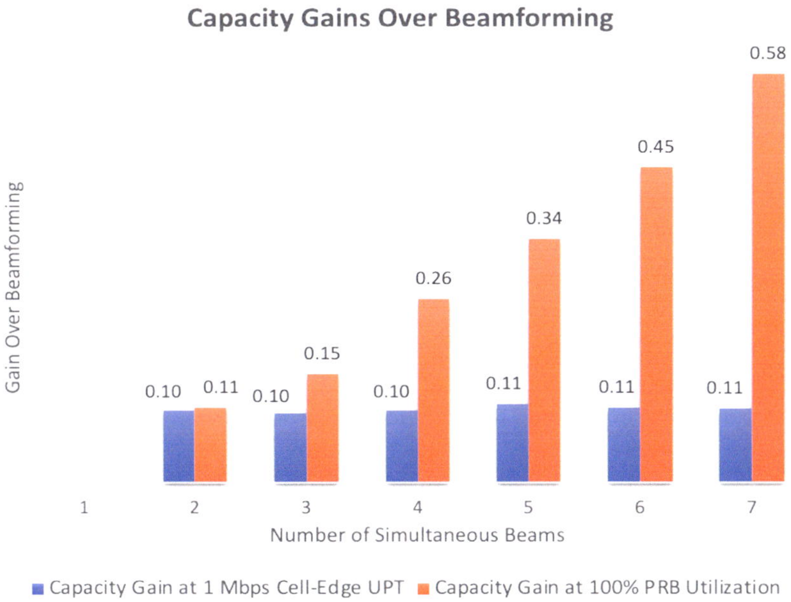

Figure 113: *Eight-element MU-MIMO capacity gains over beamforming using different KPIs.*

Although the amount of network traffic at 100% PRB utilization increases linearly with the number of simultaneous active beams, the amount of network traffic that can support 1 Mbps 5th percentile cell-edge UPT reaches its maximum at just two simultaneous beams. It is not incorrect for someone to state that MU-MIMO increases capacity by 60% over beamforming for an eight-element adaptive array. However, no MNO would operate a network at 100% PRB utilization,[70] which is one scenario where 60% capacity improvement is realized. Perhaps there are other scenarios where MU-MIMO increases capacity 60% or more. Without the ability to perform analysis, there is no way for an operator to verify those claims or to determine the capacity gains of MU-MIMO in scenarios that MNOs consider relevant.

70 Recall that network utilization rarely exceeds 30% even during the busiest hour of the day due to the imbalance of sector traffic.

7.4.2 16-Element Adaptive Array MU-MIMO

We now analyze the 16-element MU-MIMO configuration which is capable of forming 13 beams (see Figure 101) to determine the impact of additional antenna elements to the effectiveness of MU-MIMO. Figure 114 indicates that MU-MIMO is more effective with 16 antenna elements compared to the eight-element configuration.

The second and third buffers provide the greatest impact. The second buffer provides a 29% improvement in capacity (705.6 Mbps vs

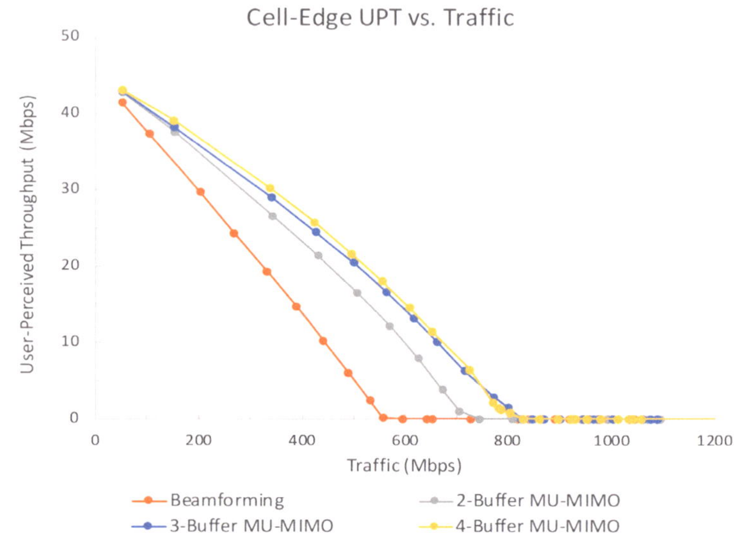

Figure 114: *Cell-edge UPT versus network traffic for 16-element MU-MIMO and simple beamforming.*

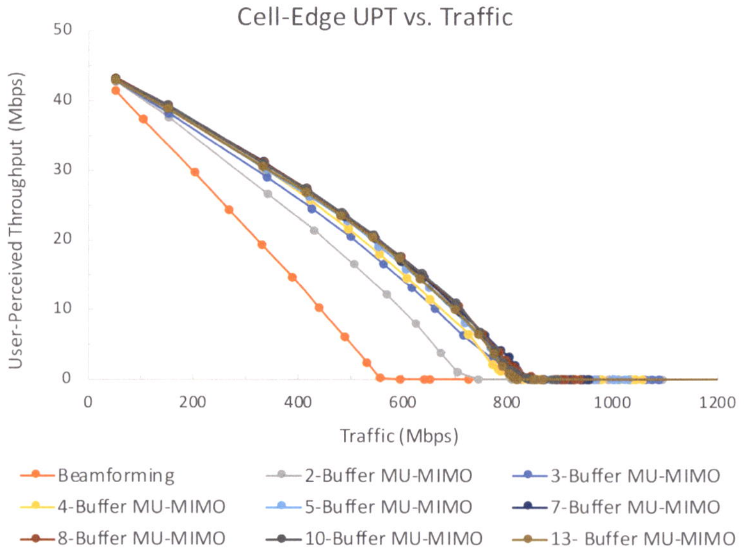

Figure 115: *Cell-edge UPT versus network traffic for 16-element MU-MIMO and simple beamforming continued.*

546.8 Mbps), while the third buffer increases capacity to 811.6 Mbps, a 48% capacity gain over simple beamforming. The fourth buffer adds little to network capacity or UPT.

To complete the analysis, Figure 115 shows the performance curves for other MU-MIMO configurations including the one where all 13 beams can be active simultaneously.

Although there are some benefits for supporting more than three simultaneous beams, it would be difficult to justify if it involves added costs. A far rosier picture can be painted by using the amount of network traffic at 100% PRB utilization instead of using the amount of network traffic to support a 1 Mbps 5th percentile cell-edge UPT.

Similar to the analysis of the eight-element adaptive array, the figure below shows the capacity gain of the 16-element MU-MIMO AAS at 100% PRB utilization and at the 1 Mbps cell-edge UPT constraint.

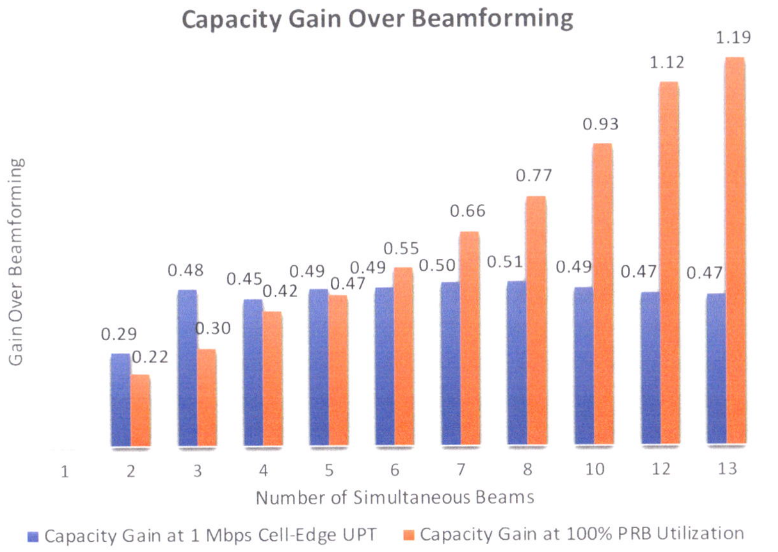

Figure 116: *Sixteen-element MU-MIMO capacity gains over beamforming using different KPIs.*

Capacity gains fade after three simultaneous beams if the 1 Mbps cell-edge UPT constraint is respected while each additional beam adds 9.2% to capacity at 100% network utilization.

7.4.3 Four-Element Adaptive Array MU-MIMO

Finally, we analyze the four-element MU-MIMO configuration. The four-element adaptive antenna can only support three beams, so it's no burden to investigate the one-buffer (i.e. beamforming), two-buffer and three-buffer configurations all at once. (See Figure 117.)

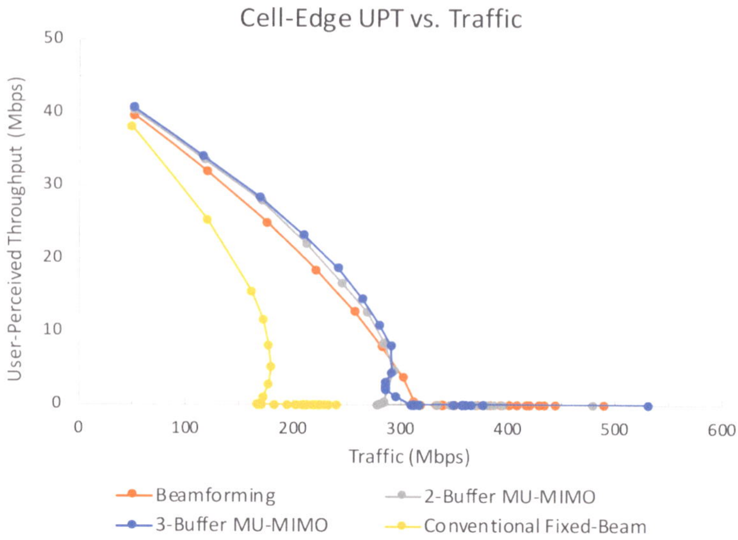

Figure 117: *Cell-edge UPT versus network traffic for four-element MU-MIMO.*

Unlike the eight-element and 16-element configurations, supporting more than one stream confers no benefit for the four-element AAS. The four-element simple beamforming configuration does provide a substantial benefit over the conventional network using fixed-beam antennas (310.3 Mbps vs. 172.1 Mbps at the 1 Mbps cell-edge UPT).

The two-buffer and three-buffer configurations display some of the aberrant characteristics seen in the conventional fixed-beam network. It is not advisable to operate the two-buffer network above 275 Mbps and the three-buffer network above 284 Mbps else the cell-edge UPT may rapidly deteriorate. (See section 7.3 for a more detailed explanation.) With simple beamforming there is less risk in trading off UPT for higher traffic.

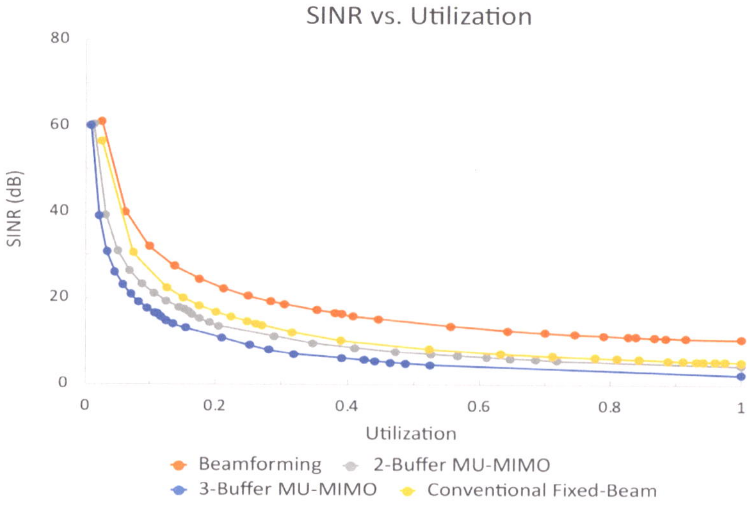

Figure 118: *SINR versus network utilization for various four-element MU-MIMO antenna configurations.*

The figure below shows how fast SINR decreases with additional beams.

As an interesting side note, observe that the network with the two-buffer MU-MIMO antenna configuration has the same average SINR as the network with the fixed-beam antennas at 100% network utilization. Given the SINRs are similar, it is no surprise the two-buffer MU-MIMO network supports twice as much traffic at 100% network utilization (479.3 Mbps vs. 239.9 Mbps) since it transmits twice as many PRBs as the conventional-antenna network.

7.5 Conclusions

Adaptive arrays promise to increase capacity and UPT over conventional fixed-beam antennas. However, caution should be exercised when deploying MU-MIMO as the capacity gains over simple beamforming may be much less than anticipated. MU-MIMO generates a substantial number of PRBs, but it may not be prudent to fill all of them with data if acceptable UPT is important.

Only a tiny aspect of AAS deployment strategy has been discussed in this chapter. For example, CoMP techniques could be used to enhance capacity and UPT. A multi-stage approach could be investigated where MU-MIMO is deployed only at busy cell sites. Various tapering techniques can be used to reduce side lobes at the expense of larger beamwidths. In this chapter, performance was limited only by co-channel interference. However, beamforming is also used to increase signal strength which can be used to overcome the impact of thermal noise in coverage-limited scenarios. As previously mentioned, since our analysis was only performed in the two-dimensional horizontal or azimuthal plane, it is possible that we are missing potential performance gains by ignoring beamforming in the third dimension or elevation plane, especially for dense urban environments.

Antenna arrays are destined to play a large role in 5G and in future air-interface technologies. Without the aid of computational techniques, it would be impossible for operators to analyze strategies to optimally deploy them.

Concluding Remarks

The appeal of using heuristics lies in its simplicity and focus on actual data. However, there are hidden assumptions that cause the heuristic method to produce puzzling predictions if it is applied over longer time frames (more than five years). Perhaps the method can be modified to address this issue, but it will never be able to predict the impacts of new technological advances. Rather than relying exclusively on data, I recommend a paradigm change by adopting an abstract model relying on a minimal number of concepts. The first has to do with the development of a stochastic algorithm which solves a set of interconnected capacity equations in scenarios that will defeat ordinary deterministic approaches. The most important equations are path loss = $10\gamma\log(d)$ and $\rho = \rho_c\,\theta$ (traffic equals the product of capacity and utilization). The second concept is based on minimizing the number of simulations by developing a technique that determines how and when the adjustment of various parameters will affect network KPIs. The most important equations are again path loss = $10\gamma\log(d)$ and, in scenarios where the wireless network is not exclusively co-channel interference limited, minSS = $SNR_{min} \times \eta$.

In the case of mobile networks where ubiquitous coverage is a requirement, performance should be limited by co-channel interference if capacity is to be maximized. This does not mean that there should be efforts to increase co-channel interference. Network configuration should be designed to minimize co-channel interference given the geographical distribution of traffic. Once this is achieved, the addition of thermal noise will only decrease performance. The added benefit of networks limited by co-channel interference is the ease of analysis versus networks where thermal noise plays a role.

In my attempt to make the book more accessible, no appeals were made to probabilistic concepts or calculus in the development of the stochastic method, despite the fact it used a stochastic process. Some statistical concepts were used to justify the use of the stochastic method and to display the shortcomings of using a simple heuristic approach. Although no formal proof was provided on the validity of the stochastic algorithm—other than results that matched the deterministic approach in scenarios where it was able to provide an answer—this is not an obstacle for its use. If using the algorithm provides insights that would not have been discovered otherwise, then does it really matter that it lacks the rigour associated with other methods? The lack of an analytic proof explaining how turbo codes approach the Shannon limit did not prevent its adoption. The alternatives to using the stochastic approach do not appear attractive.

Some may criticize the configuration of the base stations in the networks I used to demonstrate the concepts in this book. They would say that a base station configuration that resembles the hexagonal pattern displayed in Figure 13 or a Poisson point process would provide more credible results. However, there is nothing stopping them from using my methods to analyze those more-realistic network configurations. I'm just providing a method and some case studies demonstrating its efficacy.

I could have written the book describing the algorithm in more general terms, but I wanted the reader to avoid the pitfalls that plagued my development of the model. There is nothing preventing the reader from modifying the algorithm, as I mentioned that I've developed several variants that gave similar results. What intrigued me was that there was nothing available other than naïve inspection of OMs or full-blown Monte Carlo simulators, which I consider inappropriate for network dimensioning. Now there is a third option.

To demonstrate the power of using a model, we will redo the network-planning exercise of predicting how many cell sites are required to support 40% per annum traffic growth over a 10-year period. (See Chapter 1.) To perform this operation manually as specified by the heuristic method would require the planning engineer to add approximately $1.4^{10} - 1 = 28$ times the current number of cell sites. If we assume that the amount of available spectrum will double in the next 10 years, he or she will only need to manually add 14 times the current number of cell sites. Of course, the planning engineer would not know these numbers in advance, and in fact the actual number of new cell sites will depend on whether the traffic distribution changes over time which it will if one uses the heuristic planning method. Using the stochastic method to determine the number of new cell sites, this model-based approach will simply use the formula 1.4^{years}, since there is no reason to assume that the distribution of traffic changes over time. Additionally, the stochastic method allows the planning engineer to determine how long-term plans are altered if new wireless technologies are introduced. For example, if small cells are introduced and the power of the macro cells are reduced by 10 dB, Table 41 shows that the capacity of the network will increase seven-fold (946.7 Mbps versus 135.8 Mbps) if on average five small cells are placed in each macro cell. This means the number of macro cells need only double if the amount of spectrum doubles. Alternatively, if 5G and MU-MIMO will be used exclusively in 10 years, then a 16-element MU-MIMO antenna supporting at least three simultaneous streams will increase capacity six-fold (811.6 Mbps versus 135.8 Mbps), which means slightly more than double the current number of macro cells are required given a doubling in spectrum holdings.

As promised in the Introduction, I will use the concepts developed in this book to test the validity of the statements:

- The sole purpose of acquiring spectrum is to increase network capacity.
    - False. Additional spectrum does increase capacity, and it is not its sole purpose. Additional spectrum also increases user-perceived throughput to support higher speed applications. (Chapter 4)
- Adding cell sites always increase network capacity.
    - False. Adding macro cell sites to a small-cell network can reduce network capacity. (Chapter 6)
- More complicated wireless technology features are always more valuable than simple ones.
    - False: Carrier Aggregation is much simpler to deploy than CoMP, yet it provides nearly the same amount of capacity increase without sacrificing user-perceived throughput. (Chapter 4)
- Wireless network planning is very time consuming and difficult.
    - False. Sophisticated network planning tools can be entered into a compact spreadsheet that would barely fill this page. (Chapters 1 and 5)
- The lower path-loss exponent, the better the network performance.
    - False. Lower path-loss exponent decreases network capacity in interference-limited scenarios. (Chapter 3)
- Network capacity is greatly influenced by the design of the scheduler used by the base-station radio.
    - False. Sophisticated scheduler designs do not make any impact on the performance of actual wireless networks. They are chiefly used to meet arbitrary performance targets established by standards-development organizations. (Chapter 2)
- Wireless Priority Service is a very expensive feature to implement for the mobile network operator. (MNO)
    - False. Wireless Priority Service costs next to nothing to implement and does not adversely impact network KPIs. (Chapter 2)
- Low-frequency spectrum bands, due to their lower path-loss properties, are always more valuable than high-frequency spectrum bands.
    - False: In dense urban areas, if the MNO only has high-frequency spectrum, it may be to its economic advantage to deploy more of the same high-band spectrum than to deploy a low-frequency band it does not already possess. (Chapter 5)

- Every macro cell sector has traffic hotspots, so it makes sense to find those hotspots and place a small cell to offload traffic.
    - False. It is better to deploy a small cell network in areas of high traffic density with no macro overlay or one that is attenuated. (Chapter 6)
- A network using MU-MIMO (multi-user multiple-input and multiple-output) —which is an advanced antenna system capable of serving multiple subscribers simultaneously—will always provide higher network capacity than a network using simple beamforming antennas which only can transmit data to one subscriber at a time.
    - False. Unless the adaptive array has a sufficient number of elements, a MU-MIMO antenna with the same number of elements as a simple beamformer may provide less network capacity due to higher co-channel interference. (Chapter 7)
- It is always true that when radios in a wireless network transmit more often, the amount of wireless traffic will be greater.
    - False. With the absence of cell-specific reference symbols, 5G NR will reach its maximum capacity at less than 50% utilization if standard omni-directional antennas are used. (Chapter 3)

I could have provided many more examples of false statements that appear intuitively correct, but I mention only those addressed in the book.

Most academic papers in the telecom industry model the benefits of the latest wireless technology feature. Back in the early '90s, CDMA was the popular technology everyone was analyzing, followed by MIMO in the late '90s, OFDMA in the 2000s, and now Massive MIMO. What doesn't get as much attention is the methodology that models the impacts each innovation brings. Fortunately, the methodology required to test each new technology doesn't have to change. The method I presented works equally well for CDMA, OFDMA, TDMA and other digital air-interface technologies.

I have only covered a fraction of the problems I have solved using these techniques. The following are topics I will save for perhaps another book:

- Uplink analysis
- Small data-packet analysis
    - Useful for Internet of Things
- Frequency reuse analysis
    - Is universal frequency reuse better than a traditional frequency reuse plan (FDMA)?
- The impacts of various versions of Cooperative Multipoint

(CoMP) and enhanced Inter-Cell Interference Coordination (eICIC)

- Fixed wireless access (FWA)
  - How does using a directional antenna at the CPE affect network performance?
  - When are indoor CPEs with omni-directional antennas suitable for FWA?
- Massive MIMO versus adaptive beamforming.

Of the case studies presented in this book, I have only provided the highlights. To obtain additional insights, you must start programming, or at least start playing with numbers. Most of the benefits will come from solving problems using the techniques outlined in this book. Only then will you see how *C/I*, thermal noise, traffic and other factors influence network performance. Each simulation should be regarded as an experiment. You can tweak various parameters and observe how they affect important KPIs. You don't need to rely on some expert working for a telecom equipment vendor to determine the efficacy of a new feature, or some article on the Internet on how to deploy a new technology. In the age when more decisions are influenced by computer algorithms, there is no excuse not to use this power, especially when hundreds of millions, if not billions, of dollars are at stake.

Detractors will often attack someone's analysis because it did not account for some aspect of the air-interface standard. The criticism is that the analysis is too abstract and there is a possibility that the ignored feature might affect critical aspects of the model, thus invalidating any results arising from it. Most mathematicians are aware of the quote attributed to statistician George Box: "Essentially, all models are wrong, but some are useful." He has also said:

> *Since all models are wrong the scientist cannot obtain a "correct" one by excessive elaboration. On the contrary following William of Occam he should seek an economical description of natural phenomena. Just as the ability to devise simple but evocative models is the signature of the great scientist so overelaboration and overparameterization is often the mark of mediocrity.* [71]

Hopefully, the examples in this book have justified the use of abstraction and models in the analysis of wireless networks.

Those who still think that computer-based approaches are superfluous should take heed of another quote from John Tukey, a mathematician whose ideas we've encountered throughout this book.

71 George Box, "Science and statistics", *Journal of the American Statistical Association*, 71: 791–799, 1976.

> *In a world in which the price of calculation continues to decrease rapidly, but the price of theorem proving continues to hold steady or increase, elementary economics indicates that we ought to spend a larger and larger fraction of our time on calculation.*

I regret that I didn't appreciate what Tukey knew many decades earlier, when he penned the above quote.

This book can be read in two ways. The majority will read the text and gain some insight on how the telecom world works, especially from the viewpoint of an MNO. The second group are readers who need to evaluate the properties and effectiveness of numerous telecom strategies. These readers must put into practice the algorithms described in this book. Much like other skills, evaluating telecom strategies cannot be learned from just reading a book. It requires practice. If reading is all it takes to be a stock trader, musician, entrepreneur, athlete or published author, I suspect more people would be making a living in those roles. This book gives you the tools to perform wireless telecom analysis. It is up to you to learn it.

Glossary

θ	link utilization
ρ	traffic (typically measured in Mbps)
η	adjusted thermal noise floor
γ	radio frequency path-loss exponent
τ	distance scaling factor
ρ_c	capacity of the link
3G	3rd Generation (A mobile communications system of the 3rd generation)
3GPP	3rd Generation Partnership Project
4G	4th Generation (A mobile communications system of the 4th generation)
5G	5th Generation (A mobile communications system of the 5th generation)
AAS	Advanced Antenna Systems
ABS	Almost Blank Subframe
AMPS	Advanced Mobile Phone Service
ASFN	Adaptive Single Frequency Network
bps	bits per second
C/I	carrier-to-interference ratio
CA	Carrier Aggregation
CAPEX	capital expenditure
CDMA	code division multiple access
CoMP	Coordinated Multipoint
COST Hata	a radio propagation model
CPE	customer-premises equipment
CRS	Cell-specific Reference Signals
dB	decibel
dBi	antenna gain against an isotropic reference antenna measured in decibels
EDA	exploratory data analysis
eICIC	enhanced Inter-cell Interference Coordination
FCFS	First Come First Served
FFT	Fast Fourier Transform
FM	frequency modulation
FSL	free-space path loss
FWA	fixed wireless access
Gb	gigabit
GB	gigabyte
HSPA	High-Speed Packet Access
Hz	hertz
kHz	kilohertz
KPI	key performance indicator

LDPC	low-density parity-check
LTE	Long Term Evolution
LTE-A	Long Term Evolution Advanced
MATLAB	Matrix Laboratory – a proprietary programming language
Mb	megabit
MB	megabyte
Mbps	Megabits per second
MHz	megahertz
MIMO	multiple input multiple output
minSS	minimum-required signal strength
MNO	mobile network operator
MU-MIMO	multi-user multiple-input and multiple-output
mW	milliwatt
NR	5G New Radio
OFDM	orthogonal frequency-division multiplexing
OM	operational measurement
OPEX	operating expenses
PRB	Physical Resource Block
QoE	Quality of Experience
R&D	research and development
RE	Range Extension
RF	radio frequency
rnf	receiver noise floor
SINR	signal-to-interference-plus-noise ratio
SNR	signal-to-noise ratio
SNR_{min}	minimum-required signal-to-noise ratio
SU-MIMO	single-user multiple-input and multiple-output
TD-LTE	Time Division-Long Term Evolution
UE	user equipment
UPT	user-perceived throughput
W	watt
Wi-Fi	wireless local area network technology based on the IEEE 802.11 standard
WiMAX	Worldwide Interoperability for Microwave Access
WPS	Wireless Priority Service

Bibliography

K. Azarian, R. Patwardhan, C. Lott, D. Ghosh, R. Gowaikar, R. Attar, "On Spatial Load Balancing in Wide-Area Wireless Networks," *proceeding of 2012 IEEE Wireless Communications and Networking Conference (WCNC)*, pp. 2230-2235, April 2012.

H. Balakrishnan, MIT LCS, V. N. Padmanabhan, Microsoft Research, G. Fairhurst, M. Sooriyabandara, University of Aberdeen, U.K., "TCP Performance Implications of Network Path Asymmetry," *IETF RFC 3449,* December 2002.

G. Box, "Science and statistics", *Journal of the American Statistical Association*, 71: 791–799, 1976.

N. Chen and S. Jordan, "Throughput in Processor-Sharing Queues," *IEEE Transactions on Automatic Control*, vol. 52, no. 2, pp. 299–305, February 2007.

J. Dongarra and F. Sullivan, "Guest Editors Introduction to the top 10 algorithms," *Computing in Science & Engineering*, vol. 2, issue 1, pp. 22-23, January 2000.

D. C. Hoaglin, F. Mosteller and J. Tukey, *Data Tables, Trends, and Shapes*, John Wiley, New York, 1985.

H.Holma & A.Toskala, *WCDMA for UMTS: HSPA Evolution and LTE*, John Wiley & Sons, 2010.

M. K. Karray and M. Jovanovic, "A Queueing Theoretic Approach to the Dimensioning of Wireless Cellular Networks Serving Variable Bit-Rate Calls," *IEEE Transaction on Vehicular Technology*, vol. 62, no. 6, July 2013.

L. Kleinrock, *Queuing Systems, Vol 2, Computer Applications*, Wiley-Interscience, New York, NY, 1976.

L. Liu, Y. Li, B. L. Ng and Z. Pi, "Radio Resource and Interference Management for Heterogeneous Networks," *Heterogeneous Cellular Networks*, John Wiley & Sons Ltd, 2013.

D. G. Manolakis, V. K. Ingle, S. M. Kogon, *Statistical and Adaptive Signal Processing*, Artech House, 2005.

S. Ross, *Introduction to Probability Models*, Tenth Edition, Elsevier Inc., 2010.

P. Rysavy, "HSPA to LTE-Advanced: 3GPP Broadband Evolution to IMTAdvanced (4G)", *Rysavy Research*, September 2009.

J. Salo and E. Zacarías, "Analysis of LTE Radio Load and User Throughput," *International Journal of Computer Networks & Communications (IJCNC)*, vol. 9, no. 6, pp. 33-45, November 2017.

C. E. Shannon, "A Mathematical Theory of Communication," *The Bell System Technical Journal*, vol. 27, pp. 379–423, 623–656, July, October, 1948.

A. J. Viterbi, *Principles of Spread Spectrum Communication*, Addison-Wesley Publishing Company, 1995.

Index

Made in the USA
Las Vegas, NV
03 November 2026

57422209R00143